PROTEIN BIOCHEMISTRY, SYNTHESIS, STRUCTURE AND CELLULAR FUNCTIONS

ENZYMES AND ENZYME ACTIVITY

STRUCTURE, BIOLOGY AND CLINICAL SIGNIFICANCE

Protein Biochemistry, Synthesis, Structure and Cellular Functions

Additional books in this series can be found on Nova's website under the Series tab.

Additional e-books in this series can be found on Nova's website under the e-book tab.

Microbiology Research Advances

Additional books in this series can be found on Nova's website under the Series tab.

Additional e-books in this series can be found on Nova's website under the e-book tab.

PROTEIN BIOCHEMISTRY, SYNTHESIS, STRUCTURE AND CELLULAR FUNCTIONS

ENZYMES AND ENZYME ACTIVITY

STRUCTURE, BIOLOGY AND CLINICAL SIGNIFICANCE

EVA M. LASHINSKI
EDITOR

New York

Library of Congress Cataloging-in-Publication Data

ISBN: 978-1-62417-696-8

Library of Congress Control Number: 2012955727

Published by Nova Science Publishers, Inc. † New York

Contents

Preface

In this book, the authors present current research in the study of the structure, biology and clinical significance of calpain, enzymes and enzyme activity. Topics discussed include monitoring the activation of calpain I for sensitive assessment of pathogenesis and neuroprotectant action; the myeloperoxidase enzyme as a diagnostic biomarker and target for new drug developments; the biological effects of calpain inhibitors on human phagocyte functions; ALDH activity of Artemia as a tool for the investigation of the toxicity of antifouling paints; and calpain as an indicator of organo-phosphorous-induced delayed neuropathy.

Chapter I – Myeloperoxidase (MPO, EC 1.11.2.2) is an iron-containing peroxidase enzyme released primarily by neutrophils and, to a lesser extent, by monocytes and some tissue macrophages. The MPO unique ability to catalyze the hydrogen peroxide (H_2O_2) dependent oxidation of chloride (Cl^-) to hypochlorite (ClO^-) – the most reactive oxidant produced by neutrophils – accounts for its primary function as a bactericidal enzyme linked to the innate host defense. Nevertheless, its relevance in the pathophysiology of numerous diseases has become evident only in recent years. For example, circulating levels of MPO have been shown to predict risks for major adverse cardiac events, whereas the levels of MPO-derived oxidants serve as specific biomarkers for disease progression. In addition to cardiovascular diseases, many clinical and experimental studies have implicated MPO in cancer, neurodegenerative disorders or lung diseases, among others. In this chapter, MPO structure and complex catalytic cycle, as well as the intrinsic role in health and diseases are reviewed. A concise overview on the future perspectives of this enzyme as a diagnostic biomarker and target for new drug developments is also provided.

Chapter II – Calpain is a calcium-sensitive protease whose over-activation has been linked to a variety of degenerative conditions. This chapter will focus on calpain's involvement in neurodegeneration, however its pathogenic role is not restricted to the nervous system. It has been implicated in cardiac pathology, kidney hypoxia, respiratory disorders, myopathy, bone erosion, and eye diseases. Of the many substrates that calpain cleaves in brain and peripheral tissues, the cytoskeletal protein αII-spectrin has been identified as being useful to measure pathogenic calpain activation during cellular toxicity and certain diseases. Calpain-mediated spectrin breakdown has been used extensively to monitor cytoskeletal damage, one of the sensitive precursors of neuronal pathology found in animal models of excitotoxic insults and human brain injury. Cleavage of spectrin by calpain generates a distinct set of large breakdown products (BDPs) that are unusually stable. Detection and measurements of BDPs have allowed calpain-mediated proteolysis to be established as an early event contributing to the neurodegeneration associated with ischemia, hypoxia, genetic defects, vasospasm, traumatic brain injury, and seizures, and it may be involved in brain aging and Alzheimer-type pathogenesis. Being sensitive markers of excitotoxicity and other types of cellular deterioration, the spectrin fragments make up an important assay of pathology using specific labeling by antibodies against the cleavage site recognized by calpain. The ability to detect the earliest signs of pathology in experimentally-induced models of disease states allows for the careful screening of potential therapeutic agents. Especially in brain tissue targeted by acute, definable neuronal insults, neuroprotectants can be assessed for their ability to reduce calpain-mediated cytoskeletal damage and associated damage to central synapses. Excitotoxic calpain activation is commonly associated with reductions in synaptic markers, particularly among pyramidal neurons of the hippocampus. In fact, calpain-mediated spectrin breakdown levels have been found to correspond with synaptic deterioration as well as with subsequent cell death in the hippocampal CA1 subfield. The excitotoxic hippocampal slice model is often utilized for neuroprotectant characterization and screening, in which slice cultures are briefly exposed to an excitotoxin, and the neuronal stimulation subsequently quenched with glutamate receptor antagonists. Potential neuroprotectants, applied to the slices before or after the insult, are tested for effects on cytoskeletal proteolysis and synaptic decline evident 2-48 hours later. Monitoring spectrin breakdown and synaptic markers in *in vitro* and *in vivo* disease models is a valuable first step in identifying lead compounds for therapeutic indications. They are also sensitive measures for preclinical toxicology testing.

Chapter III – The intracellular proteases have several functions in regulating various cellular systems. Among them, proteases that are activated by calcium (i.e. calpain), which is a second messenger in cell regulation. Proteases duties consist mainly in the degradation of cytoskeletal proteins, modification of steroid receptors and activation of protein kinases. It has been established that there are at least two types of calpain. One type requires calcium concentrations in the order of mM for their activity, while the other type requires calcium concentrations in the order of μM. Organophosphorus (OPs) compounds, which include pesticides, industrial chemicals and some medicines, are the principal class of insecticides and chemical warfare nerve agents. They have become the most widely used insecticides in the world since the 1970s. Clinically, the symptoms of OP poisoning are divided into four categories: acute cholinergic syndrome, intermediate syndrome, chronic organophosphorus-induced neuropsychiatric disorders (COPIND) and organophosphorus-induced delayed neuropathy (OPIDN). OPIDN has been observed in more than 70,000 human cases in the last 75 years and it has been known since the 1930s when tuberculosis patients were treated with phosphor-creosote, an uncharacterised mixture of esters derived from phosphoric acid and coal-tar phenols. In 1930, thousands of paralysis cases occurred in the United States due to the consumption of Jamaica ginger extract contaminated with tri-ortho-cresyl phosphate (TOCP). Later, similar epidemics were reported in Morocco, Netherlands, Yugoslavia, France, South Africa, Sri Lanka and India. OPIDN is a neurodegenerative disorder characterised by ataxia progressing to paralysis with a concomitant central and peripheral distal axonopathy, and this neuropathology occurs similar to Wallerian-type degeneration. Initially, neuropathy target esterase (NTE) phosphorylation is involved and then an imbalance in free calcium homeostasis occurs, which leads to the activation of calpain. This sequence of events promotes protein digestion in the terminal axon. Calpain activation has been used to identify Wallerian-type axonal degeneration and this biomarker has been related with the appearance of lesions in long tracts of central and peripheral nervous system. These lesions have caused signs and symptoms that include tingling of the hands and feet. This tingling is followed by sensory loss, progressive muscle weakness and flaccidity of the distal skeletal muscles of the lower and upper extremities and ataxia, which appear about 8-14 days after poisoning. Therefore, calpain evaluation became an excellent tool to indicate axonopathies that affects people around the word.

Chapter IV – The intracellular proteases have several functions in regulating various cellular systems. Among them, proteases that are activated by calcium (i.e. calpain), which is a second messenger in cell regulation. Proteases duties consist mainly in the degradation of cytoskeletal proteins, modification of steroid receptors and activation of protein kinases. It has been established that there are at least two types of calpain. One type requires calcium concentrations in the order of mM for their activity, while the other type requires calcium concentrations in the order of μM. Organophosphorus (OPs) compounds, which include pesticides, industrial chemicals and some medicines, are the principal class of insecticides and chemical warfare nerve agents. They have become the most widely used insecticides in the world since the 1970s. Clinically, the symptoms of OP poisoning are divided into four categories: acute cholinergic syndrome, intermediate syndrome, chronic organophosphorus-induced neuropsychiatric disorders (COPIND) and organophosphorus-induced delayed neuropathy (OPIDN). OPIDN has been observed in more than 70,000 human cases in the last 75 years and it has been known since the 1930s when tuberculosis patients were treated with phosphor-creosote, an uncharacterised mixture of esters derived from phosphoric acid and coal-tar phenols. In 1930, thousands of paralysis cases occurred in the United States due to the consumption of Jamaica ginger extract contaminated with tri-ortho-cresyl phosphate (TOCP). Later, similar epidemics were reported in Morocco, Netherlands, Yugoslavia, France, South Africa, Sri Lanka and India. OPIDN is a neurodegenerative disorder character-ised by ataxia progressing to paralysis with a concomitant central and peripheral distal axonopathy, and this neuropathology occurs similar to Wallerian-type degeneration. Initially, neuropathy target esterase (NTE) phosphorylation is involved and then an imbalance in free calcium homeo-stasis occurs, which leads to the activation of calpain. This sequence of events promotes protein digestion in the terminal axon. Calpain activation has been used to identify Wallerian-type axonal degeneration and this biomarker has been related with the appearance of lesions in long tracts of central and peripheral nervous system. These lesions have caused signs and symptoms that include tingling of the hands and feet. This tingling is followed by sensory loss, progressive muscle weakness and flaccidity of the distal skeletal muscles of the lower and upper extremities and ataxia, which appear about 8-14 days after poisoning. Therefore, calpain evaluation became an excellent tool to indicate axonopathies that affects people around the word.

Chapter V – Calpain has been implicated as a regulator of actin cytoskeleton, cell migration, apoptosis, and inflammation, and calpain is a potential therapeutic target for various disorders, including neurodegenerative and inflammatory diseases.

Calpain inhibitors, including peptide aldehydes and α-mercapto-acrylic acid derivatives, are widely used as a valuable tool to elucidate the physiological and pathological roles of calpain. The biological effects of potent calpain inhibitors on human phagocytes (neutrophils and monocytes) are unique, and the inhibitors induce active migration (chemotaxis) of human phagocytes and delay neutrophil apoptosis with rapid activation of the signaling pathways, including Rac, mitogen-activated protein kinases, and phosphatidylinositol 3-kinase. Potent calpain inhibitors have been shown to activate human formyl peptide receptors and/or other G-protein coupled receptors in addition to inhibition of the calpain activity, which may be responsible for the characteristic effects of calpain inhibitors on human phagocyte functions.

Chapter VI – Leaching of toxic substances from the matrix of antifouling paints affects not only the fouling organisms but also "non-target" biota. *Artemia* nauplii have been found to be suitable test organisms for measurement of acute toxicity effects of certain antifouling paints. The present study addresses the impact of a self polishing copper antifouling paint on the activity of the brine shrimp aldehyde dehydrogenase (ALDH). Total ALDH activities from nauplii exposed to several sublethal concentrations ranging from $(S/V)_{50}$ to $1/4(S/V)_{50}$ were determined. Decreased enzymatic activities were observed in all antifouling paint concentrations tested, ranging from 2.4% inhibition at $1/4(S/V)_{50}$ to 53.9% at $(S/V)_{50}$. The results of this preliminary investigation suggest that ALDH activity of *Artemia* nauplii could be used as a biomarker for the evaluation of toxic activity of antifouling paint on non target organisms.

In: Enzymes and Enzyme Activity
Editor: Eva M. Lashinski

ISBN: 978-1-62417-696-8

Chapter I

Myeloperoxidase - A Paradoxical Enzyme - From Host Defense to Disease

***María Díaz-González*[1], *Antonio Francesko*[2], *Margarida M. Fernandes*[2] *and Tzanko Tzanov*[*2]**

[1]Instituto de Microelectrónica de Barcelona, IMB-CNM (CSIC)
[2]Group of Molecular and Industrial Biotechnology, Department of Chemical Engineering, Universitat Politècnica de Catalunya, Passeig Ernest LLuch / Rambla Sant Nebridi s/n, Terrassa, Spain

Abstract

Myeloperoxidase (MPO, EC 1.11.2.2) is an iron-containing peroxidase enzyme released primarily by neutrophils and, to a lesser extent, by monocytes and some tissue macrophages. The MPO unique ability to catalyze the hydrogen peroxide (H_2O_2) dependent oxidation of

[*] Corresponding author: Group of Molecular and Industrial Biotechnology, Department of Chemical Engineering, Universitat Politècnica de Catalunya, Passeig Ernest LLuch / Rambla Sant Nebridi s/n, 08222 Terrassa, Spain

chloride (Cl^-) to hypochlorite (ClO^-) – the most reactive oxidant produced by neutrophils – accounts for its primary function as a bactericidal enzyme linked to the innate host defense. Nevertheless, its relevance in the pathophysiology of numerous diseases has become evident only in recent years. For example, circulating levels of MPO have been shown to predict risks for major adverse cardiac events, whereas the levels of MPO-derived oxidants serve as specific biomarkers for disease progression. In addition to cardiovascular diseases, many clinical and experimental studies have implicated MPO in cancer, neurodegenerative disorders or lung diseases, among others. In this chapter, MPO structure and complex catalytic cycle, as well as the intrinsic role in health and diseases are reviewed. A concise overview on the future perspectives of this enzyme as a diagnostic biomarker and target for new drug developments is also provided.

1. Introduction

Heme peroxidases, containing the prosthetic heme group, are the most abundant peroxidases in all living forms. This kind of enzymes catalyzes one- and two-electron oxidation reactions of a great diversity of inorganic and organic compounds with hydrogen peroxide (H_2O_2). Heme peroxidases can be classified into two large superfamilies: plant, fungal, and bacterial peroxidases and mammalian and other animal peroxidases. Both superfamilies arose independently, thus their primary and tertiary structures and even the nature of the prosthetic group differ greatly.

The denomination mammalian peroxidase superfamily derived from the fact that its members, myeloperoxidase (MPO), eosinophil peroxidase (EPO), lactoperoxidase (LPO), and thyroid peroxidase (TPO) were originally identified in mammals.

MPO first isolated by Agner in 1941 [1], is the most prominent representative of the mammalian peroxidase superfamily. MPO is one of the most abundant proteins (5%) [2] in human polymorphonuclear leukocytes (PMNs, also called neutrophils) and to a lesser extent it is related to monocytes [3] and tissue macrophages [4]. Neutrophils are the most abundant type of leukocytes in the body, making up 70 % of all white blood cells. Neutrophils are the first cell type in human beings that are activated in the host immune defense against infection. In neutrophils MPO is stored in the azurophilic granules and is released in response to leukocyte activation [5].

There are three major isoforms of human MPO designated MPO I, MPO II, and MPO III [6]. These forms exhibit differences in molecular weight, cationicity, hydrophobicity, sensitivity to inhibitors, and localization within the azurophil granule subpopulations.

For a long time, MPO main function has been considered to be killing pathogens through the generation of reactive oxygen species (ROS). However, recent evidence has extended this view by demonstrating that MPO is also an important factor in the initiation and progression of various inflammatory diseases.

2. MPO Structure

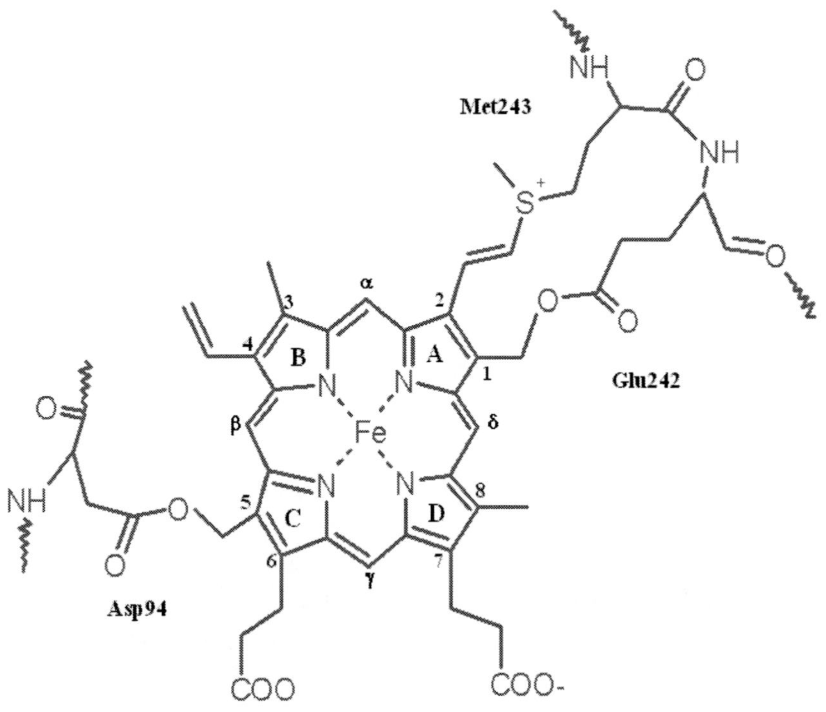

Figure 1. Ferri-protoporphyrin IX in MPO.

Structurally, MPO is the best characterized enzyme of the mammalian peroxidases. MPO [7] (first structure elucidate in 1992) is a strongly cationic 146 kDa dimer with a single disulfide bridge between symmetry-related halves

(73 kDa). Each half molecule contains two polypeptides of 14.5 and 58.5 kDa, the latter being glycosylated. The small polypeptide is composed of 106 and the large of 467 amino acids. There are five additional intra-chain disulfides in the large polypeptide and one in the small. Each half of dimeric MPO contains one iron, present as covalently bound heme and one calcium. The calcium binding site has typical pentagonal bipyramidal coordination. However, as it has been revealed by Fiedler *et al.* in 2000 [8], an uncommon way of heme binding is found in human MPO. The prosthetic group in mammalian peroxidases is a ferri-protoporphyrin IX (Figure 1). This heme group in mammalian peroxidases is joined with the apoprotein by two ester linkages. The modification of the 1- and 5-methyl groups on pyrrole rings A and C of the heme group (Figure 1) enabling the formation of ester linkages with the carboxyl groups of a conserved aspartate (Asp94) and glutamate (Glu242) is the most important structural feature of mammalian peroxidases. In addition to that only in MPO the β-carbon of the vinyl group on pyrrole ring A form a covalent bond with the sulfur atom of methionine 243, giving rise to a sulfonium ion linkage (Figure 1). This threefold linkage of heme is unique compared to other heme proteins and makes the porphyrin ring slightly curved. The uncommon way of heme binding is responsible for the peculiar spectroscopic features of MPO (red shift of the Soret band to 430 nm) and also for its special redox behavior discussed in the following section [9].

3. MPO Catalytic Cycle

A simplified mechanism of MPO catalytic cycle is shown in Figure 2. H_2O_2 reacts with the native form of MPO (Fe^{3+}) to generate the two-electron oxidised species compound I (a ferryl π-cation radical species where an oxygen atom is coupled by a double bond to the iron). Compound I may be converted back to the ferric enzyme via two-electron reduction by (pseudo) halides (halogenation cycle), or via two sequential one-electron reduction reactions via formation of compound II (peroxidase cycle). An additional intermediate, compound III (oxyperoxidase), can be generated via reaction of the ferric form with superoxide radicals, from ferrous form with O_2, or from compound II with H_2O_2. Compound III is unreactive towards most substrates, and is usually a catalytic 'dead-end'; this specie will however slowly oxidize ascorbate and paracetamol (acetaminophen) to radicals, and is implicated in superoxide dependent catalytic activities of MPO [10].

One particularity of MPO is the midpoint potential of the Fe^{3+}/Fe^{2+} couple, which is +25 mV [11]. This value is much more positive than that of other heme peroxidases (-180 to -300 mV) and similar to that of globins (myoglobin and hemoglobin), which have to be stable in the ferrous form to function as reversible oxygen carrier. Contrary to this redox behavior, the midpoint potentials for heme peroxidases guarantee a stable ferric form (necessary for H_2O_2 reduction). Similar to its singular spectroscopic features these special redox properties have their origin in the electron-withdrawing sulfonium linkage and heme distortion [12]. As a consequence ferrous MPO must be taken into account for the enzymatic mechanisms and makes the enzyme thermodynamically competent to oxidize chloride. As can be seen in Figure 2 ferrous form can react either with H_2O_2 to generate compound II or with dioxygen to compound III.

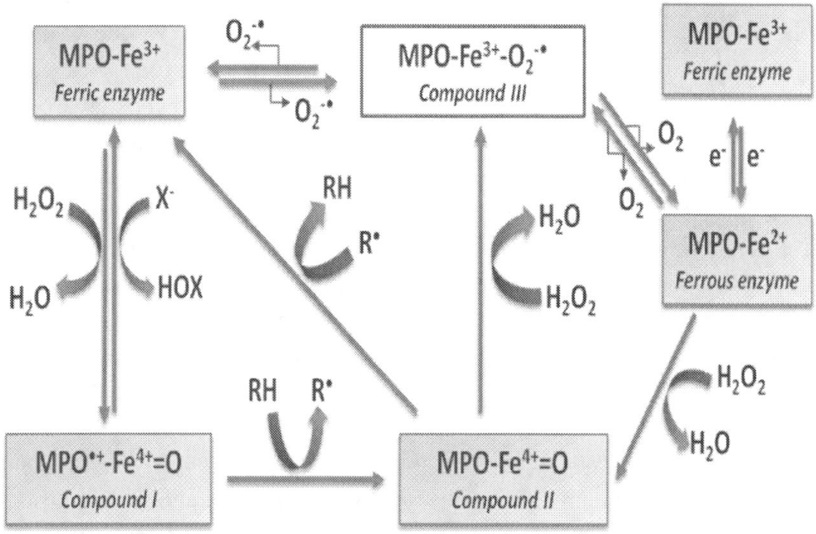

Figure 2. MPO catalytic cycle.

Due to the high reduction potentials of the compound I/native enzyme (1.16 V), compound I/compound II (1.35 V) and compound II/native enzyme (0.97 V) couples, MPO can oxidize multiple substrates via its halogenation and peroxidase cycles [13,14]. The catalytic activity of MPO is therefore partitioned between the two cycles via competition between peroxidase substrates and (pseudo) halides for compound I. MPO activity largely depends further on the pH of the medium [15].

3.1. MPO Substrates

The key substrate of MPO is H_2O_2, produced by neutrophils in response to various stimuli from molecular oxygen via the NADPH oxidase ''respiratory burst'' complex. Most of the H_2O_2 is consumed by MPO to generate compound I. MPO also reacts with superoxide anion radicals to produce compound III which is involved in the hydroxylation of aromatic substrates [10, 16].

3.1.1. Substrates in the Halogenating Cycle of MPO

MPO catalyzes the oxidation of halide (Cl^-, Br^-, I^-) ions and the pseudo-halide thiocyanate (SCN^-) by H_2O_2, producing oxidizing and halogenating agents. In aqueous media main product of halide oxidation is a HOX/OX^- (X= Cl^-, Br^-, I^- or SCN^-) mixture [17]. MPO halogenating substrate preference has been the subject of different studies, yielding contradictory results. In general, it is assumed that Cl^- is the preference substrate for MPO due to its high concentration in plasma (100-140 mM) compared to those for I^- (< 1 μM), Br^- (20-100 μM) and SCN^- (20-120 μM). Some studies, however, have revealed that SCN^- is an equally preferred substrate under physiological conditions and is oxidized to hypothiocyanite at both neutral and acidic conditions [18]. Anyway, HClO is the most powerful bactericidal oxidant produced by neutrophils in appreciable amounts. It is approximately 100–1000 times more toxic than superoxide anion radicals and H_2O_2 and many species of bacteria are killed readily by this acid [19]. Hypothiocyanite being considerably less reactive than HClO may also play an important role in host defense as it can divert H_2O_2 away from the production of HClO limiting bacterial killing and tissue damage.

It is difficult to establish the ratio between the different MPO halogenation products since it dependents on the substrate availability and the local pH, and thus is highly dynamic within the biological context of an inflammatory response. As far the chlorination reaction, there is no well-defined pH optimum for HClO production, because it depends on the relative concentrations of Cl^- and H_2O_2 [15, 20]. At low pH chloride acts as a competitive inhibitor for H_2O_2 while at higher pH, H_2O_2 inhibits the chlorination reaction. In general, the chlorinating activity of MPO decreases with increasing pH and at neutral pH the enzyme catalyzes mainly reactions of the peroxidase cycle. Thus the presence of this kind of substrates also affects the production of HClO by MPO. Regarding the effect of substrate availability on MPO halogenation products, a recent study by Talib *et al.* [21] proved that

MPO mediated damage differed in smokers and non-smokers due to the elevated levels of plasma SCN^- present in smokers [22]. An increased HOSCN formation by MPO diminishes the damage to methionine, tryptophan, lysine, histidine, and tyrosine residues in proteins implicated in multiple human inflammatory diseases. The differences between these two oxidant systems have recently been reviewed by Pattison *et al.* [23]. Briefly, as it has been mentioned above, HOSCN is less reactive than other hypohalous acids, especially HClO [24]. HClO reacts rapidly with sulfur and nitrogen atoms including those present in thiols, thioethers, amines, and amides. Through cysteine residue oxidation this acid can activate the (inactive) pro- forms of matrix metalloproteinases.

Reaction with amines produces chloramines, which are liposoluble and more toxic than HClO itself. Nearly all of the common α-amino acids serve as substrates for the acid to produce reactive aldehyde [25]. HClO also reacts with nucleobases [26], unsaturated lipids and choleresterol [27].

HClO reacts with H_2O_2 generating singlet oxygen, with anion radicals or Fe(II) to form hydroxyl radicals, or with nitrite to yield the powerful chlorinating and nitrating compound NO_2Cl. As it will be explained below, all products of MPO halogenating cycle have antimicrobial and antiviral activity, but may also damage host tissues and contribute to inflammatory tissue injury.

3.1.2. Substrates for the Peroxidase Cycle of MPO

Many different inorganic and organic compounds are known to be peroxidase substrates for MPO. During this cycle substrates are oxidized in a two-step one-electron process by compound I and compound II. Due to the different redox potentials of these two MPO intermediates, some compounds can be oxidised by compound I, and not (or poorly) by compound II resulting in "trapping" MPO at this point in its enzymatic cycle. This property has been exploited for inhibition of MPO halogenation cycle. Peroxidase substrates include the amino acids tyrosine [28], ascorbate, ureate, or nitrite [29, 30] as well as xenobiotics [31] and drugs [32]. Similar to MPO halogenating products the radicals generated via the metabolism of peroxidase substrates may damage biomolecules like lipids or proteins [24].

4. MPO in Health

A number of studies have described the important role for MPO in the microbicidal activity of neutrophils [33] through the generation of ROS. The inactive form of the enzyme is stored in the azurophilic granules of neutrophils, involved in both innate and acquired immune response, and activated upon phagocytosis of bacteria [34]. Neutrophils, normally found in the bloodstream, move toward the affected area during inflammation by a process known as chemotaxis. The antimicrobial activity of neutrophils is associated with the NADPH oxidase activation [35] and with the degranulation process [36], which releases ROS and the granule components, including MPO, into the phagosome.

Once fully assembled at the cytosolic face of the plasma membrane the NADPH oxidase complex transfers electrons from NADPH in the cytosol to oxygen in the phagosome to form superoxide (Figure 3). Subsequent reactions lead to the formation of other species, including H_2O_2. Hydrogen peroxide, like superoxide, has low reactivity and is not lethal to microorganisms, unless present at high concentrations (mM).

Microbicidal and cytotoxic activities of H_2O_2 are due to secondary products such as HClO, HOSCN or $RO^·$ produced by MPO. Moreover, HClO generation represents only the first step in production of many bioreactive (i.e. hydroxyl radical, chloroamines, etc.) molecules that vary in their half-life, diffusing capacity beyond the phagosome, and reactivity with potential targets [24].

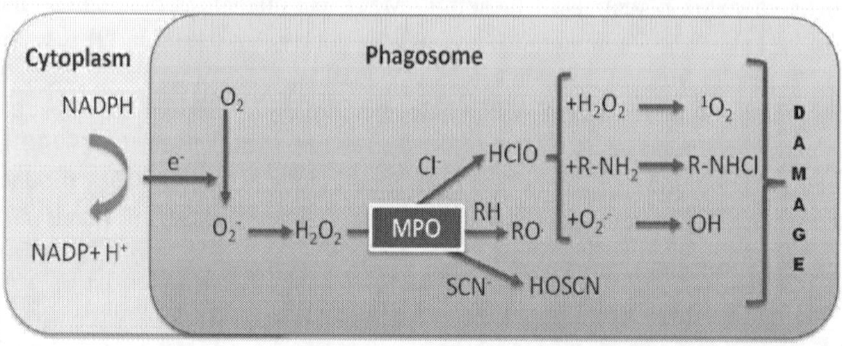

Figure 3. MPO-dependent mechanism of bacterial killing in neutrophils.

Although this described MPO-dependent mechanism is the predominant mode of bacterial killing in neutrophils, partial or complete MPO-deficient individuals do not appear to be considerably more susceptible to infections, revealing that MPO-independent mechanisms are involved in microbicidal activity of neutrophils as well [5, 37, 38].

The biological significance of MPO including its implication in the pathophysiology of numerous diseases will be discussed in the following section.

5. MPO Related Diseases

Regulated expression of MPO gene transcription is critical to maintain the constant optimal levels of the enzyme and derived oxidants in order to control acute infections. However, in a range of immune and inflammatory disorders altered systemic levels of the enzyme have been identified. Recent insights in the interactions between MPO and the derived oxidants with their biological substrates revealed the related biological damage in many of these conditions. Modulated immune responses derived from MPO non-catalytic effects and proinflammatory events of MPO-derived oxidants contribute to initiation and progression of tissue damage [39]. In this context, both deficiency and overexpression of MPO has been linked to the tissue injury caused by its direct or indirect inflammatory reactions (Figure 4). Infections, carcinogenesis, renal diseases, atherosclerosis, neurodegenerative disorders and other chronic inflammatory conditions are some examples briefly discussed in this section.

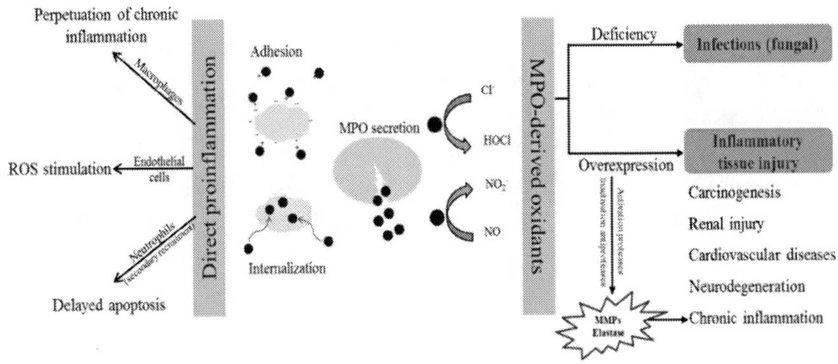

Figure 4. Summary of the MPO multiple pathways in pathogenesis of different disorders.

5.1. Disorders Related to MPO-Oxidation Products

5.1.1. MPO Deficiency

MPO deficiency is a disorder that occurs when individuals have abnormally low or nonexistent levels of the enzyme and predisposed immune deficiency. In most cases the condition is inherited, unless in cases of acquired temporary deficiency caused by other medical conditions. It is known that prevalence of the hereditary MPO deficiency is one case in 2000-4000 individuals.

Studies have demonstrated an increased susceptibility to severe infections of the subjects suffering the deficiency, due to the microbicidal defect in neutrophils lacking MPO [40]. The most frequently studied complications related to MPO-deficiency were those of pneumonia [41] and polymicrobial sepsis [42].

Nevertheless, the MPO-deficient neutrophils are also adapted to the long-term absence of the enzyme, as they show slightly higher antimicrobial potential compared to inhibitor-treated neutrophils [43]. This implicates other MPO-independent antimicrobial mechanisms to compensate for the lack of the enzyme and presumably accounts for the fact that nearly half of the MPO-deficient individuals are usually detected with the typical symptoms of infections [44].

On the other hand, the number of infections also depends on the degree of MPO-deficiency (total *vs* partial) and exposure to the microbial challenge (high *vs* low burden infections) [45]. The enzyme defect was further associated with a selective impairment in the killing of different type of microorganisms [46]. Indeed, the variation in susceptibility of different pathogens to defected oxygen-dependent antimicrobial systems has been confirmed in experiments on MPO-deficient animals. Killing fungi largely depends on the MPO presence (fungal infections are the most frequent in MPO-deficient patients), whereas bacteria differ in their susceptibility to MPO-deficient neutrophils [47, 48].

Nevertheless, animal studies do not necessarily reflect the MPO-deficiency in humans as the disorder causes distinct antimicrobial activity alterations primarily due to the different MPO levels in the neutrophils in different organisms. For example, whereas the studies on MPO-deficient neutrophils in mice and humans match in the high susceptibility to *Candida albicans* infection, the results on the clearance of *Staphylococcus aureus* were contradictory [49, 50].

Another clinical manifestation of MPO deficiency involves an increased incidence of malignant tumors, although contradictory reports could be found on this topic. Whereas some authors relate a defective MPO system and neutrophil-mediated tumor cell cytotoxicity [50], others found no evidences for increased incidence of cancer among the MPO-deficient individuals [40].

A recent investigation linked various cancer groups with low MPO concentrations in neutrophils [51]. The lowest MPO levels were found in the cancer patients additionally suffering fungal infections.

5.1.2. MPO Overexpression

MPO in Cancer

Continuous oxidative stress in the site of inflammation results in oxidative damage of genomic DNA of proliferating cells, and cause mutations and development of cancer [52]. As MPO-derived oxidants together with other ROS have genotoxic effect, they can induce carcinogenesis by modulation of the genes that regulate cell proliferation or by DNA mutations [53]. In line with this, inflammatory response in the lungs caused after the exposure to initiators such as asbestos or smoking is associated with the initiation and progression of lung carcinogenesis *in vivo*, with high concentrations of MPO detected [54]. Oxidative and nitrosative stresses have also been identified in many other metastatic environments [55].

A number of studies have reported the influence of genetically determined MPO levels on cancer risk. These works focus on several polymorphisms in the MPO promoter regions that affect MPO systemic levels, relating them to cancer development. For example, increased MPO expression have been related to acute myeloid leukemia [56] and adenocarcinoma [57], whereas lower MPO transcription rates has been associated with a decreased risk of larynx cancer [58], bladder cancer [59] and rare liver neoplasm [60]. Although these findings are consistent with the overall role of MPO in carcinogenesis, some controversies exist considering the risks for developing a specific cancer.

MPO in Renal Injury

An elevated MPO activity has been found in many renal diseases [61]. Experimental studies were performed on animals to determine the damage derived from MPO oxidants [62]. By inducing perfusions on the kidney, a model for renal ischemia/reperfusion injury in sepsis and after kidney transplantation could be developed. These studies confirmed in most cases the abundance of HClO in the site of injury and related damage to surrounding cells and tissues. Continuous migration of neutrophils and MPO into the injured kidney is consistent with a contribution of MPO-oxidants to further kidney damage [63, 64]. On the other hand, evidences from experimental models of nephritis show that MPO mediates glomerular injury after migration of neutrophils to the region [65]. In both cases the presence of H_2O_2 in high counts following the injury is further compromising the treatment options.

Recently, several functional polymorphisms of MPO and other host response genes have been implicated in pathogenesis and treatment of both acute and chronic renal failure [66, 67]. A two-step genetic study provided evidences that four polymorphisms across the length of the MPO gene with systemic markers of oxidative stress and a common corresponding haplotype influence the risk of adverse outcomes in hospitalized adults with or at risk for acute kidney injury [68]. The effect seems to be mediated by modulation of the extracellular release of MPO.

MPO in Inflammatory Vascular Disease

The detection of MPO and specific markers of MPO-catalyzed oxidation in human atherosclerotic lesions is consistent with the involvement of MPO in atherosclerosis development *in vivo* [69]. MPO-generated oxidants affect the cholesterol and lipid metabolism to promote atherosclerosis. Both low-density and high-density lipoproteins undergo oxidative modifications, thereby inducing cholesterol deposition and loss of anti-inflammatory lipoprotein, respectively. Lipoproteins isolated from atherosclerotic lesions contain several-fold higher levels of tyrosine nitrated residues compared to circulating lipoproteins from healthy donors [70, 71]. Moreover, the stabilization of the lesion is promoted in the presence of active forms of extracellular matrix metalloproteinases-(MMPs) [72]. In turn, the MMPs activities are also modulated by MPO-derived oxidants. HClO has been shown to activate MMPs by oxidation of the thiol moiety in the active center of the enzyme (mechanism termed "cysteine switch") [73], allowing for destabilization of atherosclerotic plaque [74]. Paradoxically, the proteolytic hydrolysis during inflammation is strictly regulated by the MPO level in the site of inflammation, as high concentrations of HClO will induce inactivation of naturally occurring tissue inhibitors of MMPs (TIMPs) [75]. Despite the fact that not all experimental animal models were in agreement with the above described pathogenesis, majority of these models confirmed MPO-derived oxidation as a key event for the initiation and progression of the disease [76, 77]. These results are in a good agreement with the recent studies on asymptomatic individuals for the prediction of cardiovascular events risk, in which the plasma MPO concentrations provided a reliable prognostic value for the prediction of long-term cardiovascular risk [78-80].

MPO and Neurodegeneration

Many studies pointed out an important role of oxidative stress in pathogenesis of neurodegenerative diseases. Several markers of oxidative stress are increased in brain tissue of patients with Alzheimer's disease and traces of lipid peroxidation found in cerebral cortex and brain neurons suggest a pathogenic role of catalytically active MPO in the disease [81]. Moreover, MPO has been also detected in microglia/macrophages in and around the lesions in multiple sclerosis, in contrast to a non-affected brain tissue [82, 83]. On the other hand, leucocytes of the affected individuals were found to be MPO-deficient. These data suggest that the enzyme levels and derived oxidation certainly play an important role in the propagation of multiple sclerosis, especially as emerging evidences show that similar neuro-degenerative conditions are tightly connected with the lower antioxidant capacity in the organism [84, 85].

Regarding MPO polymorphism, recent experiments in transgenic mice suggest that there is a relation between the aberrant MPO expression on one, and specific pattern of phospholipid peroxidation and neuronal dysfunction on the other side, contributing to development of defect in Alzheimer's disease [86]. However, as in the case of the overall cancer risk, studies relating the polymorphisms were not always in agreement regarding the specific risks in different gender and geographical cohorts [87, 88].

MPO in Other Chronic Inflammatory Disorders

Implication of MPO-derived oxidants in many chronic tissue injuries has been also described. Among others, a contributory role of oxidative stress has been proposed in the development of lung inflammation and rheumatoid arthritis. In both cases large number of active neutrophils could be detected or directly increased levels of MPO were measured [89, 90]. MPO driven modulation of oxidative stress has been also confirmed in animal models with lung injuries [91]. On the other hand, the proinflammatory milieu with synergistic effects of neutrophils-derived oxidants and proteases accounts for the development of joint injury in rheumatoid arthritis [92, 93]. Such example could be also found in other chronic inflammatory conditions. It is known that both elevated levels of proteases and excessive neutrophils characterize chronic wound environments [94, 95]. As it has been pointed in inflammatory vascular disease discussion, one of the major factors governing the protease-antiprotease balance that allows healing (typical for acute wounds) is the MPO-derived generation of HClO. The excess of HClO reacts with most biological molecules [96], including protease inhibitors [97], thereby promoting the proteolytic damage of healthy tissues. However, besides protease inhibitors MPO also directly attenuates the elevated proteases activities thereby protecting the host from the tissue degradation [98]. Paradoxically, in healing processes some proteases contribute to the tissue reconstruction by controlled degradation of fibrotic proteins. In addition, a recent study has demonstrated that the same proteases are also employed by the host to counteract bacterial infections [99]. Remains unclear which of the two substrates (protease vs antiprotease) will be oxidized *in vivo*, most probably depending on the concentrations of all components from this complex system at the wound site. The preferred substrate would further determine whether the healing will be initiated or the chronicity will be prolonged. Such unpredictable behavior of MPO and derived oxidants makes the treatment of chronic inflammatory diseases extremely difficult.

5.2. Direct Inflammatory Properties of MPO

MPO has been recently shown to exert proinflammatory effects beyond its oxidative properties and in many cases independent of its enzymatic activity. These effects are expressed after the enzyme interaction with various cells. Modulation of the cellular functions occurs through MPO adhesion onto cells, MPO internalization by cells, as well as the MPO-induced cell activation and migration. For example, earlier studies suggested that interactions between MPO and macrophages enhance the antimicrobial properties of these cells [100, 101]. However, more recent studies revealed that in this way activated macrophages also perpetuate inflammation in a range of pathologies including arthritis [102] and lung inflammation [103]. MPO has also been shown to adhere on leukocytes, acting as an autocrine modulator of cell function independent of its enzymatic activity [104]. The enzyme adhesion thus activates the neutrophils and delays apoptosis, thereby contributing to chronic inflammation [105]. A recent study demonstrated that the principal mechanism for MPO adhesion onto neutrophils in their recruitment is driven by physical forces between the positively charged enzyme and negatively charged extracellular glycoproteins on the surface of the cell [106]. In this secondary neutrophils recruitment (neutrophil-secreted products that modulate recruitment of neutrophils [107]), MPO evokes highly directed cell locomotion solely dependent on electrostatic interactions with the cell surface. In vivo, neutrophils recruitment was shown to be MPO-dependent in a model of hepatic ischemia and reperfusion, upon intraportal delivery of the protein and in the cremaster muscle exposed to local inflammation or to intraarterial MPO application. The authors envisioned the neutrophil migration through a reverse mechanism in which MPO itself reduces the electrostatic interactions.

Besides in neutrophils, exposure to MPO has been found to modulate cellular functions in various other cell types. MPO is readily internalized in human umbilical vein endothelial cells, inducing secretion of interleukins and stimulating generation of reactive oxygen species [108]. The internalization of MPO in endothelial cells also affects the vasoregulatory system [109]. A direct proinflammatory effect of MPO on lung epithelial cells has also been demonstrated after the enzyme internalization [110], increasing heme-oxygenase-1 (HO-1) expression and inducing DNA damage. Ambiguously, increased stimulation of HO-1 enzyme also allows the cell to counteract HClO-mediated mitochondrial dysfunction and cell death [111]. The increased HO-1 gene expression may thus represent a critical adaptive response to maintain endothelial cell viability at sites of persistent inflammation.

6. Therapeutic Implications

MPO and its derived oxidant species play an important role in the pathophysiology of many diseases characterized by acute and chronic inflammation. MPO participate in a range of events that contribute to the initiation and propagation of the inflammatory response and, as such, is involved in tissue pathology [39]. These effects suggest that MPO-targeted strategies may be of therapeutic benefit. In fact, several approaches use MPO as a predictor of diseases (diagnostic biomarker) or as a target for the development of new drugs. In the following section, a concise review on the state-of-the-art and future perspectives of MPO in this field is provided.

6.1. MPO as a Biomarker

According to the National Institutes of Health (NIH), biomarker is defined as a “characteristic that is objectively measured and evaluated as an indicator of normal biological processes, pathogenic processes, or pharmacological responses to a therapeutic intervention". The biomarker responds to the presence and progression of inflammatory disease in an individual, being both over or underexpressed in response to an inflammatory process.

MPO has been regarded as a novel biomarker with significant research and commercial interest [112]. As it has been explained in previous sections the overexpression of this enzyme has been associated with normal tissue damage thus contributing to several disorders such as cardiovascular diseases [113], carcinogenesis [56, 114, 115], renal [116] and lung injury [117], atherosclerosis [118], multiple sclerosis [119], Alzheimer's disease [87], brain infarction [120], Parkinson's disease [121] and chronic wounds [95].

Of particular recent interest is the proposed involvement of MPO in the development of cardiovascular diseases, considered a major threat to global health. In the last decade, clinical trials have shown evidences that circulating levels of MPO can be a strong predictor of adverse cardiovascular events in medical practice and extend the prognostic information gained from traditional biomarkers [122]. In fact, MPO has been shown to be a strong predictor in patients with acute coronary syndrome [122, 123], chronic heart failure [124], and acute myocardial infarction [125]. The first study addressing this issue assessed the MPO level in patients with chest pain and concluded that a single initial measurement of plasma MPO independently predicts the early risk of myocardial infarction and risk of major adverse cardiac events in the ensuing 30-day and 6-month periods [123]. More recently, Heslop *et al.* [126] reported that elevated MPO levels more than doubled the risk for cardiovascular mortality over a 13-year period, and measuring both MPO and CRP (C-reactive protein) provided added benefit for risk prediction to improve long-term risk assessment and CVD patient outcomes.

The advantage of MPO over other biomarkers, in the setting of heart failure and acute coronary syndrome, is the early increase of MPO concentration in response to the acute event [127, 128]. Traditional cardiac markers such as troponin I or T, myoglobin, creatine kinase MB isoform, and C-reactive protein levels, are present only in acute event, after myocardial damage/necrosis have occurred [129]. Evidences that inflammatory activation by MPO precedes the onset of overt cardiovascular disease (CVD) by many years, set MPO as a potent biomarker/predictor of future coronary artery disease in a healthy population [130].

MPO has also been reported as an intracellular marker in acute myeloid leukemia diagnosis. Immunohistochemical staining of MPO to demonstrate that the leukemic cells are derived from the myeloid lineage was reported to diagnose acute myeloid lineage [131]. Moreover, MPO staining has been revealed as an important tool in the diagnosis of myeloid sarcoma, contrasting with the negative staining of lymphomas, which can otherwise have a similar appearance [132]. Another condition with clinical relevance and associated with the overexpression of MPO is the chronic wounds. The prolonged inflammation in chronic wounds cause the accumulation of neutrophils at wound site where they undergo apoptosis and release their constituents such as MPO [95]. MPO overexpression appears to have diagnostic and prognostic value and is associated to several conditions/diseases with clinical and economic impact. Therefore, finding assays that monitor MPO levels and activity, accurately and fast, has captured much attention in the past few years [39]. To meet this growing demand, the development of portable, rapid and low cost biosensing devices detecting MPO in serum and wound exudate have experienced great development.

Immunoassays are the most commonly used technique to asses MPO levels and have been widely reported [133-138]. Among the available immunoassays, lateral flow immunoassays (LFI) are the main detection techniques used in biological fluids because of their pretreatment simplicity and high ability to treat many samples in a short time [139]. LFIs are essentially immunoassays adapted to operate along a single axis to suit the test strip format. The sample binds to its specific antibody and travel along an absorbent surface layer until it encounters a detection zone. Further reaction with appropriate substrate result in a colour readout [139].

Using this basic principle, a recent study developed a method to detect MPO using an avian high-affinity recombinant antibody [137]. After isolation, the MPO-specific recombinant antibody was incorporated into a rapid simple and sensitive one step LFI. The final detection was a result of the specific enzymatic activity of MPO (Figure 5). This one-step feature of the developed assay was made possible by the antibody strong affinity for MPO, obviating the need for "sandwich signal enhancement" steps.

Electronic-based point-of care (POC) devices have also gaining ground in this field [138, 140, 141]. An iridium oxide (IrOx) nanowire- based device was proven to be useful to detect both CRP and MPO inflammatory proteins. An immunoassay was built onto the iridium oxide nanowires and each binding event led to a specific electrical parameter perturbations [140]. The development of a more simple strip-based assay for amperometric detection of MPO was also achieved by Windmiller *et al.* [141]. The biosensor utilizes tetramethyl benzidine (TMB) as a redox mediator to enable high sensitivity quantification of physiological levels of MPO.

Due to the increased incidence of chronic wounds and related healthcare costs, POC devices able to detect wound fluid biomarkers have also been central to several investigations. Currently, the clinicians often visually inspect the wound to evaluate the infection status or rely on time-consuming and costly ex-vivo analyses requiring special equipment. Among the most recent advances are remote monitoring devices evaluating wound-related parameters (margin, volume, depth and area) [142] or off-site diagnostic kits requiring painful wound fluid collection [143]. A device able to determine the healing phase of a wound accurate and inexpensively has been recently developed.

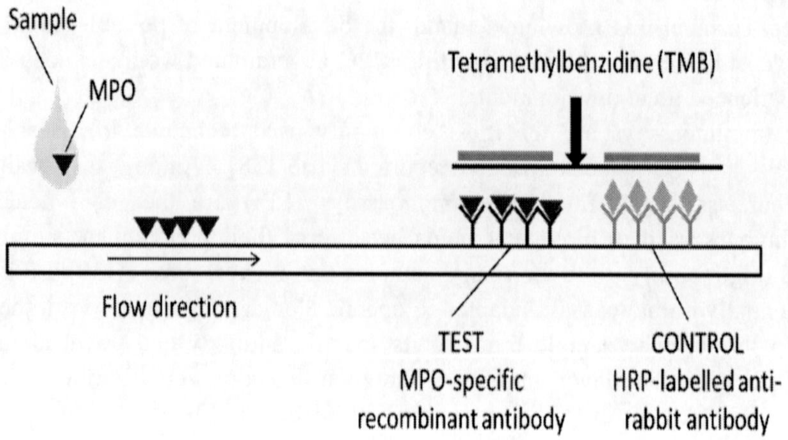

Figure 5. Schematic representation of lateral flow immunoassay applied on studies from [137]. The MPO-specific recombinant antibody binds to MPO present in the sample that further reacts with its substrate 3,3',5,5'–tetramethyl benzidine (TMB) at the test site to develop a blue colour. The control comprises an IgG antibody labelled with HRP that also reacts with TMB to develop blue colour.

Garcia *et al* [144] reported a wound diagnosis device comprising a surface with agents that are specific and bind to a desired biomarker, including MPO. Similar technology was developed by Gubitz *et al* [145] in which enzymes present in wound fluid, such as MPO, react with specific chromogenic substrates and lead to a colour change within minutes.

The development of such devices should feature fast and simple techniques for MPO detection leading to a timely therapeutic intervention of CVD (which will ultimately lead to increased survival rates) and efficient wound management.

6.2. Target for Drug Development (Inhibitors of MPO)

MPO plays an important role in the innate host defense generating strong oxidants as previously explained. However, an excessive or misplaced production of these compounds is also responsible of the tissue damage associated with many inflammatory diseases. Thus, a considerable interest is concentrated on the development of MPO-specific inhibitors. At first glance, the development of these MPO-specific inhibitors for *in vivo* use is potentially difficult, given the beneficial role of the MPO system in pathogen killing. However, the rather large variation of MPO levels in the healthy population and the relatively mild phenotype of fully or partially MPO-deficient individuals suggest that a therapeutic window must exist in which MPO levels can be maintained for host defense.

A number of different strategies targeting different paths in the MPO catalytic cycle have been investigated. One approach is the prevention of MPO-derived oxidants accumulation by decreasing the availability of H_2O_2 [146]. Among the compounds studied, diphenylene iodonium chloride (DPI) has been the most extensively investigated. This compound has also the ability to inhibit other sources of $O_2^{\bullet-}$ formation and consequently the H_2O_2 levels [147]. Another possibility is to block the access of H_2O_2 to its binding site in the enzyme (i.e. salicyl and benzylhydroxamic acids) [148].

Small anions, including azide and cyanide, can also be effective MPO inhibitors. Acting as "heme poisons" they bind competitively to Fe^{2+} at MPO heme centre in place of its typical halogenation substrates. The uses of these agents is however limited by their multitude of interactions with other heme centres and subsequent toxicity [96].

Other compounds that avidly bind MPO have also been investigated as potential inhibitors. Caeruplasmin, a copper containing plasma protein, is reported to inhibit MPO peroxidase and halogenation activities [149] whereas the anticoagulant heparin, that bind electrostatically to MPO, lead to its release from the artery wall thus diverting the oxidation from critical sites [150]. "Suicide inhibitors" (i.e. hidrazinas and hydrazidas) that react with the heme prosthetic group of MPO after their own enzymatic oxidation has also been reported [151] as irreversible enzyme inhibitors [152, 153]. Scavengers of the MPO produced hypohalous acids, especially those for HClO have also been proposed to limit the MPO-mediated damage [154]. However, these strategies show some important limitations (lack of specificity, high concentrations or toxicity) and the attention has been focused on inhibit MPO by diverting the enzyme from the halogenation cycle.

As the catalytic activity of MPO is partitioned between the halogenation and peroxidase cycles via competition between peroxidase and (pseudo) halides substrates, the use of competitive amounts of peroxidase substrates can be used to avoid the pathway of the enzyme halogenating activity. Thiol-containing molecules were reported to be efficient on the inhibition of HClO production by MPO [155, 156]. In fact, previous results from our group have shown that thiolated chitosan induce the inhibition of MPO in a dose dependent manner, being more effective for chitosan with higher thiolation grade [156].

Phenols and plant polyphenol extracts have also been shown to act as efficient MPO inhibitors [157]. Plant polyphenolic extracts, which structures vary from simple molecules to highly polymerized compounds, are well-known for their antioxidant capacity and scavenging activity over both free radical and non-radical reactive species [158], metal-chelating capability [159] and inhibitory activity over radical-generating enzymes [160]. These characteristics set plant polyphenolic extracts as efficient inhibitors of MPO [161]. Indeed, their propensity to bind proteins presumably accounts for the fact that polyphenols inhibit virtually every enzyme tested in vitro [162].

Studies performed in our group indicated that polyphenolic extracts from *Hamamelis virginiana* efficiently controlled the activity of MPO [163]. The same extracts loaded onto collagen sponges were further released and were sufficient for complete inhibition of MPO [164]. Other studies performed with collagen, collagen/hyaluronic acid (HA) and collagen/HA/chitosan (CS) sponges loaded with epigallocatechin gallate (EGCG), catechin (CAT) and gallic acid (GA) (Figure 6) were developed and evaluated as active chronic wound dressings. All specimens containing polyphenols were found to efficiently inhibit MPO activity [165].

The ability of some peroxidase substrates to be oxidised by compound I, and not (or poorly) by compound II resulting in "trapping" MPO catalytic cycle at this point of the peroxidase cycle has also been exploited for MPO inhibition. Some anti-inflammatory drugs [166, 167], tryptophan and derivatives [168], and nitroxide radicals [169] promote the accumulation the MPO redox intermediate compound II diverting the enzyme from the halogenating cycle. Nevertheless, most of these compounds are toxic (*in vivo*) at the concentration required for MPO inhibition (*in vitro*) and to date compounds that can effectively inhibit MPO in vivo at non-toxic doses have not been identified. Furthermore, this inhibitory mechanism is likely to have limited efficacy in vivo because endogenous reductants are able to reduce compound II to native enzyme.

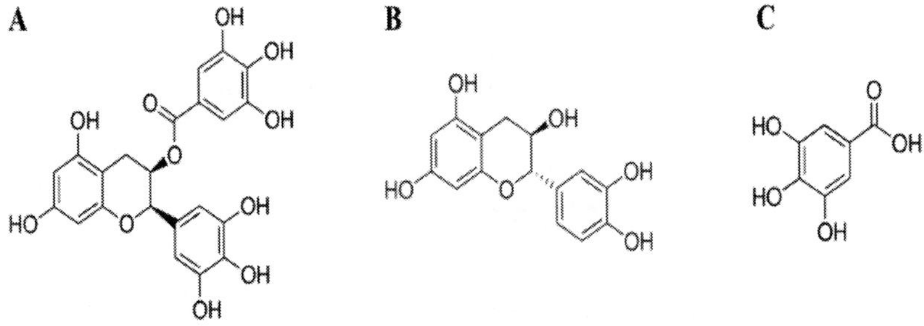

Figure 6. Structure of polyphenolic units: A) epigallocatechin gallate EGCG, B) catechin CAT and C) gallic acid GA.

Acetaminophen (paracetamol) has been recently revealed as a potential modulator of MPO activity *in vivo* [170]. Given the increased knowledge of MPO protein structure, reaction mechanisms and the vast sources that can be used, the development of new drugs that target MPO is expected to advance rapidly.

6.3. MPO Degradation of Carbon Nanotubes

MPO has been finding new and important applications on other distinct areas such as nanomaterials technology. MPO is the first and so far only human enzyme known to break down carbon nanotubes [171, 172]. The importance of this issue resides in the wide variety of applications of carbon nanotubes in clinical diagnosis, targeted drug delivery, and cancer therapy [172], and the concern of unhealthy build up in tissues. Such concern has been limiting the application of these new materials *in vivo* since the perspectives for their use in biomedical applications depend largely on their ability to degrade in the body into products that can be easily cleared out.

It has been shown that free radicals and oxidants generated in the reactions mediated by HRP and MPO can oxidize and degrade carbon nanotubes and the degradation was induced not only by their incubation with MPO but also by the incubation with neutrophils [171]. Another study provided direct evidence for the participation of MPO in the *in vivo* pulmonary oxidative biodegradation of carbon nanotubes in mice and suggest new ways to control the biopersistence of nanomaterials through genetic or pharmacological manipulations [173]. Since MPO is an abundant enzyme of inflammatory cells (neutrophils) and also effective in the oxidative biodegradation of carbon nanotubes, this discovery paves the way for the use of carbon nanotubes in inflammatory diseases where the MPO is overexpressed, allaying the concern of nanotoxicity and unhealthy accumulation of nanotubes in tissues.

Conclusion

Myeloperoxidase is a unique, multifunctional and challenging enzyme involved in both human health and disease. Traditionally, MPO has been considered the main source of reactive oxygen species for bacterial killing, and thus an important part of host defense. However, the extent to which MPO contributes to human innate immune response remains unclear, due to the uncertain immunodeficiency associated with the lack of MPO observed in mouse and human models. Recent studies have shown the relevance of this enzyme and its downstream products in the pathophysiology of numerous health disorders such as cardiovascular diseases, carcinogenesis or chronic wounds. These new findings strongly suggest MPO as a possible marker for diagnosis as well as a target for the development of new drugs. Nevertheless, additional investigations together with standardized assays are required before MPO can be routinely adopted as a biomarker in clinical practice. Further structural studies of MPO protein to provide additional insights on the reaction mechanism, are necessary in order to develop effective enzyme inhibitors *in vivo*.

References

[1] Agner, K. Verdoperoxidase. A ferment isolated from leukocytes. *Acta Physiol Scand* 1941, 2, 1–62.

[2] Schultz, J; Kaminker, K. Myeloperoxidase of the leucocyte of normal human blood. I. Content and localization. *Arch Biochem Biophys* 1962, 96, 465-467.

[3] Bos, A; Wever, R; Roos, D. Characterization and quantification of the peroxidase in human monocytes. *Biochim Biophys Acta* 1978, 525, 37−44.

[4] McMillen, TS; Heinecke, JW; LeBoeuf, RC. Expression of human myeloperoxidase by macrophages promotes atherosclerosis in mice. *Circulation* 2005,111, 2798-2804.

[5] Nauseef, WM. How human neutrophils kill and degrade microbes: an integrated view. *Immunol Rev* 2007, 219, 88–102.

[6] Miyasaki, KT; Song, JP; Rekha, A; Murthy, K. Secretion of myeloperoxidase by human neutrophils. *Anal Biochem* 1991, 193, 38-44.

[7] Zeng, J; Fennat, RE. X-ray crystal structure of canine myeloperoxidase at 3 Å resolution. *J Mol Biol* 1992, 226, 185-207.

[8] Fiedler, TJ; Davey, CA; Fenna, RE. X-ray crystal structure and characterization of halide-binding sites of human myeloperoxidase at 1.8 Å resolution. *J Biol Chem* 2000, 275, 11964–11971.

[9] Devarajan, A; Gaenko, AV; Ryde, U. Effect of covalent links on the structure, spectra, and redox properties of myeloperoxidase – a density functional study. *J Inorg Biochem* 2008, 102, 1549–1557.

[10] Kettle, AJ; Anderson, RF; Hampton, MB; Winterbourn, CC. Reactions of superoxide with myeloperoxidase. *Biochem* 2007, 46, 4888-4897.

[11] Ikeda-Saito, M; Prince, RC. The effect of chloride on the redox and EPR properties of myeloperoxidase. *J Biol Chem* 1985, 260, 8301-8305.

[12] Battistuzzi, G; Stampler, J; Bellei, M; Vlasits, J; Soudi, M; Furtmüller, PG; Obinger, C. Influence of the covalent heme–protein bonds on the redox thermodynamics of human myeloperoxidase. *Biochem* 2011, 50, 7987–7994.

[13] Arnhold, J; Furtmuèller, PG; Regelsberger, G; Obinger, C. Redox properties of the couple compound I/native enzyme of myeloperoxidase and eosinophil peroxidase. *Eur J Biochem* 2001, 268, 5142-5148.

[14] Furtmüller, PG; Arnhold, J; Jantschko, W; Pichler, H; Obinger, C. Redox properties of the couples compound I/compound II and compound II/native enzyme of human myeloperoxidase. *Biochem Bioph Res Com* 2003, 301, 551–557.

[15] Vlasova, II; Arnhold, J; Osipov, AN; Panasenko, OM. pH-Dependent regulation of myeloperoxidase activity. *Biochem* 2006, 71, 667-677.

[16] Kettle, AJ; Winterbourn, CC. Superoxide-dependent hydroxylation by myeloperoxidase. *J Biol Chem* 1994, 269, 17146-17151.

[17] Furtmüller, PG; Burner, U; Obinger, C. Reaction of myeloperoxidase compound I with chloride, bromide, iodide, and thiocyanate. *Biochem* 1998, 37, 17923-17930.

[18] van Dalen, CJ; Whitehouse, MW; Winterbourn, CC; Kettle, AJ. Thiocyanate and chloride as competing substrates for myeloperoxidase. *Biochem J* 1997, 327, 487-492.

[19] Conner, EM; Grisham, MB. Inflammation, free radicals, and antioxidants. *Nutrition* 1996, 12, 274-277.

[20] Bakkenist, AR; de Boer, JE; Plat, H; Wever, R. The halide complexes of myeloperoxidase and the mechanism of halogenations reactions. *Biochim Biophys Acta* 1980, 613, 337–348.

[21] Talib, J; Pattison, DI; Harmer, JA; Celermajer, DS; Davies, MJ. High plasma thiocyanate levels modulate protein damage induced by myeloperoxidase and perturb measurement of 3-chlorotyrosine. *Free Radical Bio Med* 2012, 53, 20–29.

[22] Morgan, PE; Pattison, DI; Talib, J;Summers, FA; Harmer, JA; Celermajer, DS; Hawkins, CL; Davies, MJ. High plasma thiocyanate levels in smokers are a key determinant of thiol oxidation induced by myeloperoxidase. *Free Radic Bio Med* 2011, 51, 1815–1822.

[23] Pattison, DI; Davies, MJ; Hawkins, CL. Reactions and reactivity of myelo- peroxidase-derived oxidants: differential biological effects of hypochlorous and hypothiocyanous acids. *Free Radic Res* 2012, 46, 975-995.

[24] Davies, MJ. Myeloperoxidase-derived oxidation: mechanisms of biological damage and its prevention. *J Clin Biochem Nutr* 2011, 48, 8–19.

[25] Hazen, SL; Hsu, FF; d'Avignon, A; Heinecke, JW. Human neutrophils employ myeloperoxidase to convert r-amino acids to a battery of reactive aldehydes: a pathway for aldehyde generation at sites of inflammation. *Biochem* 1998, 37, 6864-6873.

[26] Henderson, JP; Byun, J; Takeshita, J; Heinecke, JW. Phagocytes produce 5-chlorouracil and 5-bromouracil, two mutagenic products of myeloperoxidase, in human inflammatory tissue. *J Biol Chem* 2003, 278, 23522-235228.

[27] Ford, DA. Lipid oxidation by hypochlorous acid: chlorinated lipids in atherosclerosis and myocardial ischemia. *Clin Lipidol* 2010, 5, 835–852.

[28] Marquez, LA; Dunford, HB. Kinetics of oxidation of tyrosine and dityrosine by myeloperoxidase compounds I and II. Implications for lipoprotein peroxidation studies. *J Biol Chem* 1995, 270, 30434-40.

[29] Meotti, FC; Jameson, GNL; Turner, R; Harwood, DT; Stockwell, S; Rees, MD; Thomas, SR; Kettle, AJ. Urate as a physiological substrate for myeloperoxidase. *J Biol Chem* 2011, 286, 12901–12911.

[30] Burner, U; Furtmüller, PG; Kettle, AJ; Koppenol, WH; Obinger, C. Mechanism of reaction of myeloperoxidase with nitrite. *J Biol. Chem* 2000, 275, 20597–20601.

[31] Hofstra, AH; Uetrecht, JP. Myeloperoxidase-mediated activation of xenobiotics by human leukocytes. *Toxicology* 1993, 82, 221-242.

[32] Forbes, LV, Furtmüller, PG; Khalilova I; Turner, R; Obinger, C; Kettle, AJ. Isoniazid as a substrate and inhibitor of myeloperoxidase: Identification of amine adducts and the influence of superoxide dismutase on their formation. *Biochem Pharmacol* 2012, 84, 949-60.

[33] Klebanoff, SJ. Myeloperoxidase: friend and foe. *J Leukocyte Biol* 2005, 77, 598-625.

[34] Arnhold, J; Flemmig, J. Human myeloperoxidase in innate and acquired immunity. *Arch Biochem Biophys* 2010, 500, 92–106.

[35] Clark, RA. Activation of the neutrophil respiratory burst oxidase. *J Infec Dis* 1999,179, S309–17.

[36] Lacy, P. Mechanisms of Degranulation in Neutrophils. *Allergy Asthma Clin Immuno* 2006, 2, 98-107.

[37] Hurst, JK. What really happens in the neutrophil phagosome? *Free Radical Bio Med* 2012, 53, 508–520.

[38] Hampton, MB; Kettle, AJ; Winterbourn, CC. Inside the neutrophil phagosome: oxidants, myeloperoxidase, and bacterial killing. *Blood* 1998, 92, 3007-3017.

[39] van der Veen, BS; de Winther, MPJ; Heeringa, P. Myeloperoxidase: molecular mechanisms of action and their relevance to human health and disease. *Antioxid Red Sign* 2009, 11, 2899-2937.

[40] Kutter, D; Devaquet, P; Vanderstocken, G; Paulus, JM; Marchal, V; Gothot, A. Consequences of total and subtotal myeloperoxidase deficiency: risk or benefit ? *Acta Haematol* 2000, 104, 10-15.

[41] Koh, AY; Priebe, GP; Ray, C; Van Rooijen, N; Pier, GB. Inescapable need for neutrophils as mediators of cellular innate immunity to acute *Pseudomonas aeruginosa pneumonia. Infec Immun* 2009, 77, 5300-5310.

[42] Gaut, JP; Yeh, GC; Tran, HD; Byun, J; Henderson, JP; Richter, GM; Brennan, ML; Lusis, AJ; Belaaouaj, A; Hotchkiss, RS; Heinecke, JW. Neutrophils employ the myeloperoxidase system to generate antimicrobial brominating and chlorinating oxidants during sepsis. *PNAS* 2001, 98, 11961-11966.

[43] Klebanoff, SJ. Myeloperoxidase: contribution to the microbicidal activity of intact leukocytes. *Science* 1970, 169, 1095-1097.

[44] Kutter, D. Prevalence of myeloperoxidase deficiency: population studies using Bayer-Technicon automated hematology. *J Mol Med* 1998, 76, 669-675.

[45] Aratani, Y; Kura, F; Watanabe, H; Akagawa, H; Takano, Y; Suzuki, K; Dinauer, MC; Maeda, N; Koyama, H. Critical role of myeloperoxidase and nicotinamide adenine dinucleotide phosphate-oxidase in high-burden systemic infection of mice with Candida albicans. *J Infec Dis* 2002, 185, 1833-1837.

[46] Aratani, Y; Kura, F; Watanabe, H; Akagawa, H; Takano, Y; Suzuki, K; Maeda, N; Koyama, H. Differential host susceptibility to pulmonary infections with bacteria and fungi in mice deficient in myeloperoxidase. *J Infec Dis* 2000, 182,1276-1279.

[47] Aratani, Y; Kura, F; Watanabe, H; Akagawa, H; Takano, Y; Suzuki, K; Dinauer, MC; Maeda, N; Koyama, H. In vivo role of myeloperoxidase for the host defense. *Jpn J Infec Dis* 2004, 57,S15.

[48] Aratani, Y. Role of myeloperoxidase in the host defense against fungal infection. *Jpn J Med Mycol* 2006, 47, 195-199.

[49] Aratani, Y; Koyama, H; Nyui, S; Suzuki, K; Kura, F; Maeda, N. Severe impairment in early host defense against Candida albicans in mice deficient in myeloperoxidase. *Infec Immun* 1999, 67, 1828-1836.

[50] Lanza, F; Fietta, A; Spisani, S; Castoldi, GL; Traniello, S. Does a relationship exist between neutrophil myeloperoxidase deficiency and the occurrence of neoplasms? *J Clin Lab Immunol* 1987, 22, 175-180.

[51] Koziol-Montewka, M; Magrys, A; Paluch-Oles, J; Bogut, A; Buczynski, K; Jablonka, S. MPO and cytokines in the serum of cancer patients in the context of Candida colonization and infection. *Immunol Invest* 2006, 35, 167-179.

[52] Multhoff, G; Molls, M; Radons, J. Chronic inflammation in cancer development. *Frontiers Immunol* 2011, 2, 98.

[53] Kagan, VE; Yalowich, JC; Borisenko, GG; Tyurina, YY; Tyurin, VA; Thampatty, P; Fabisiak, JP. Mechanism-based chemopreventive strategies against etoposide-induced acute myeloid leukemia: free radical/antioxidant approach. *Mol Pharmacol* 1999, 56, 494-506.

[54] Mossman, BT. Carcinogenesis and related cell and tissue responses to asbestos: a review. *Ann Occup Hyg* 1994, 38, 617-624.

[55] Ortega, ÁL; Mena, S; Estrela, JM. Oxidative and nitrosative stress in the metastatic microenvironment. *Cancer* 2010, 2, 274-304.

[56] Reynolds, WF; Chang, E; Douer, D; Ball, ED; Kanda, V. An allelic association implicates myeloperoxidase in the etiology of acute promyelocytic leukemia. *Blood* 1997, 90, 2730-2737.

[57] Wheatley-Price, P; Asomaning, K; Reid, A; Zhai, R; Su, L; Zhou, W; Zhu, A; Ryan, DP; Christiani, DC; Liu, G. Myeloperoxidase and superoxide dismutase polymorphisms are associated with an increased risk of developing pancreatic adenocarcinoma. *Cancer* 2008, 112, 1037-1042.

[58] Cascorbi, I; Henning, S; Brockmöller, J; Gephart, J; Meisel, C; Müller, JM; Loddenkemper, R; Roots, I. Substantially reduced risk of cancer of the aerodigestive tract in subjects with variant −463A of the myeloperoxidase gene. *Cancer Res* 2000, 60, 644-649.

[59] Hung, RJ; Boffetta, P; Brennan, P; Malaveille, C; Gelatti, U; Placidi, D; Carta, A; Hautefeuille, A; Porru, S. Genetic polymorphisms of MPO, COMT, MnSOD, NQO1, interactions with environmental exposures and bladder cancer risk. *Carcinogenesis* 2004, 25, 973-978.

[60] Pakakasama, S; Chen, TTL; Frawley, W; Muller, C; Douglass, EC; Tomlinson, GE. Myeloperoxidase promotor polymorphism and risk of hepatoblastoma. *Int J Cancer* 2003, 106, 205-207.

[61] Malle, E; Buch, T; Grone, H-J. Myeloperoxidase in kidney disease. *Kidney Int* 2003 64,1956-1967.

[62] Brouwer, E; Huitema, MG; Klok, PA; de Weerd, H; Tervaert, JW; Weening, JJ; Kallenberg, CG. Antimyeloperoxidase-associated proliferative glomerulonephritis: an animal model. *J ExpMed* 1993, 177,905-914.

[63] Miura, M; Fu, X; Zhang, Q-W; Remick, DG; Fairchild, RL. Neutralization of Groα and Macrophage Inflammatory protein-2 attenuates renal ischemia/reperfusion injury. *Am J Pathol* 2001, 159, 2137-2145.

[64] Lloberas, N; Torras, J; Herrero-Fresneda, I; Cruzado, JM; Riera, M; Hurtado, I; Grinyó, JM. Postischemic renal oxidative stress induces an inflammatory response through PAF and oxidized phospholipids: prevention by antioxidant treatment. *FASEB J*, 2002, 16, 908-910.

[65] Johnson, RJ; Lovett, D; Lehrer, RI; Couser, WG; Klebanoff, SJ. Role of oxidants and proteases in glomerular injury. *Kidney Int* 1994, 45, 352-359.

[66] Jaber, BL; Pereira, BJG; Bonventre, JV; Balakrishnan, VS. Polymorphism of host response genes: Implications in the pathogenesis and treatment of acute renal failure. *Kidney Int* 2005, 67, 14-33.

[67] Doi, K; Noiri, E; Maeda, R; Nakao, A; Kobayashi, S; Tokunaga, K; Fujita, T. Functional polymorphism of the myeloperoxidase gene in hypertensive nephrosclerosis dialysis patients. *Hypertens Res* 2007, 30, 1193-1198.

[68] Perianayagam, MC; Tighiouart, H; Liangos, O; Kouznetsov, D; Wald, R; Rao, F; O'Connor, DT; Jaber, BL. Polymorphisms in the myeloperoxidase gene locus are associated with acute kidney injury-related outcomes. *Kidney Int* 2012, 82, 909-919.

[69] Daugherty, A; Dunn, JL; Rateri, DL; Heinecke, JW. Myeloperoxidase, a catalyst for lipoprotein oxidation, is expressed in human atherosclerotic lesions. *J Clin Invest* 1994, 94, 437-444.

[70] Leeuwenburgh, C; Hardy, MM; Hazen, SL; Wagner, P; Oh-ishi, S; Steinbrecher, UP; Heinecke, JW. Reactive nitrogen intermediates promote low density lipoprotein oxidation in human atherosclerotic intima. *J Biol Chem* 1997, 272, 1433-1436.

[71] Pennathur, S; Bergt, C; Shao, B; Byun, J; Kassim, SY; Singh, P; Green, PS; McDonald, TO; Brunzell, J; Chait, A; Oram, JF; O'Brien, K; Geary, RL, and Heinecke, JW. Human atherosclerotic intima and blood of patients with established coronary artery disease contain high density lipoprotein damaged by reactive nitrogen species. *J Biol Chem* 2004, 279,42977-42983.

[72] George, SJ. Tissue inhibitors of metalloproteinases and metalloproteinases in atherosclerosis. *Curr Opin Lipido* 1998, 9,413-423.

[73] Shabani, F; McNeil, J; Tippett, L. The oxidative inactivation of tissue inhibitor of metalloproteinase-1 (TIMP-1) by hypochlorous acid (HOCI) is suppressed by anti-rheumatic drugs. *Free Radical Res* 1998, 28, 115-123.

[74] Lau, D; Baldus, S. Myeloperoxidase and its contributory role in inflammatory vascular disease. *Pharmacology &amp. Therapeutics* 2006, 111, 16-26.

[75] Wang, Y; Rosen, H; Madtes, DK; Shao, B; Martin, TR; Heinecke, JW; Fu, X. Myeloperoxidase inactivates TIMP-1 by oxidizing its N-terminal cysteine residue. *J Biol Chem* 2007, 282, 31826-31834.

[76] Brennan, M-L; Wu, W; Fu, X; Shen, Z; Song, W; Frost, H; Vadseth, C; Narine, L; Lenkiewicz, E; Borchers, MT; Lusis, AJ; Lee, JJ; Lee, NA; Abu-Soud, HM; Ischiropoulos, H, and Hazen, SL. A Tale of two controversies. *J Biol Chem* 2002, 277, 17415-17427.

[77] Eiserich, JP; Baldus, S; Brennan, M-L; Ma, W; Zhang, C; Tousson, A; Castro, L; Lusis, AJ; Nauseef, WM; White, CR, and Freeman, BA. Myeloperoxidase, a leukocyte-derived vascular NO oxidase. *Science* 2002, 296, 2391-2394.

[78] Wong, ND; Gransar, H; Narula, J; Shaw, L; Moon, JH; Miranda-Peats, R; Rozanski, A; Hayes, SW; Thomson, LEJ; Friedman, JD, and Berman, DS. Myeloperoxidase, subclinical atherosclerosis, and cardiovascular disease events. *JACC: Cardiovas Imag* 2009, 2, 1093-1099.

[79] Lobbes, MBI; Kooi, ME; Lutgens, E; Ruiters, AW; Lima Passos, V; Braat, SHJG; Rousch, M; Ten Cate, H; van Engelshoven, JMA; Daemen, MJAP; Heeneman, S. Leukocyte counts, myeloperoxidase, and pregnancy-associated plasma Protein A as biomarkers for cardiovascular disease: Towards a multi-Biomarker approach. *Int J Vascular Med* 2010, 2010, 1-9.

[80] Tang, WHW; Wu, Y; Nicholls, SJ; Hazen, SL. Plasma myeloperoxidase predicts incident cardiovascular risks in stable patients undergoing medical management for coronary artery disease. *Clin Chem* 2011, 57, 33-39.

[81] Green, PS; Mendez, AJ; Jacob, JS; Crowley, JR; Growdon, W; Hyman, BT; Heinecke, JW. Neuronal expression of myeloperoxidase is increased in Alzheimer's disease. *J Neurochem* 2004, 90, 724-733.

[82] Gray, E; Thomas, TL; Betmouni, S; Scolding, N; Love, S. Elevated activity and microglial expression of myeloperoxidase in demyelinated cerebral cortex in multiple sclerosis. *Brain Pathol* 2008, 18, 86-95.

[83] Lefkowitz, DL; Lefkowitz, SS. Microglia and myeloperoxidase: A deadly partnership in neurodegenerative disease. *Free Radical Bio Med* 2008, 45, 726-731.

[84] Maes, M; Mihaylova, I; Kubera, M; Uytterhoeven, M; Vrydags, N; Bosmans, E. Increased 8-hydroxy-deoxyguanosine, a marker of oxidative damage to DNA, in major depression and myalgic encephalomyelitis / chronic fatigue syndrome. *Neuroendocrinol Lett* 2009, 30, 715-722.

[85] Chan, A; Rogers, E; Shea, TB. Dietary deficiency in folate and vitamin e under conditions of oxidative stress increases phospho-tau levels: Potentiation by ApoE4 and alleviation by S-adenosylmethionine. *J Alzheimers Dis* 2009, 17, 483-487.

[86] Maki, RA; Tyurin, VA; Lyon, RC; Hamilton, RL; DeKosky, ST; Kagan, VE; Reynolds, WF. Aberrant expression of myeloperoxidase in astrocytes promotes phospholipid oxidation and memory deficits in a mouse model of Alzheimer disease. *J Biol Chem* 2009, 284, 3158-3169.

[87] Reynolds, WF; Rhees, J; Maciejewski, D; Paladino, T; Sieburg, H; Maki, RA; Masliah, E. Myeloperoxidase polymorphism is associated with gender specific risk for Alzheimer's disease. *Exp Neurol* 1999, 155,31-41.

[88] Porrello, E; Monti, MC; Sinforiani, E; Cairati, M; Guaita, A; Montomoli, C; Govoni, S; Racchi, M. Estrogen receptor α and APOEε4 polymorphisms interact to increase risk for sporadic AD in Italian females. *Eur J Neurol* 2006, 13, 639-644.

[89] McCabe, AJ; Dowhy, M; Holm, BA; Glick, PL. Myeloperoxidase activity as a lung injury marker in the lamb model of congenital diaphragmatic hernia. *J Pediatr Surg* 2001, 36, 334-337.

[90] Stamp, LK; Khalilova, I; Tarr, JM; Senthilmohan, R; Turner, R; Haigh, RC; Winyard, PG; Kettle, AJ. Myeloperoxidase and oxidative stress in rheumatoid arthritis. *Rheumatology*, 2012.

[91] Haegens, A; van der Vliet, A; Butnor, KJ; Heintz, N; Taatjes, D; Hemenway, D; Vacek, P; Freeman, BA; Hazen, SL; Brennan, ML; Mossman, BT. Asbestos-induced lung inflammation and epithelial cell proliferation are altered in myeloperoxidase-null mice. *Cancer Res* 2005, 65, 9670-9677.

[92] Ottonello, L; Cutolo, M; Frumento, G; Arduino, N; Bertolotto, M; Mancini, M; Sottofattori, E; Dallegri, F. Synovial fluid from patients with rheumatoid arthritis inhibits neutrophil apoptosis: role of adenosine and proinflammatory cytokines. *Rheumatology* 2002, 41,1249-1260.

[93] Wright, HL; Moots, RJ; Bucknall, RC; Edwards, SW. Neutrophil function in inflammation and inflammatory diseases. *Rheumatology*, 2010, 49, 1618-1631.

[94] Trengove, NJ; Stacey, MC; Macauley, S; Bennett, N; Gibson, J; Burslem, F; Murphy, G; Schultz, G. Analysis of the acute and chronic wound environments: the role of proteases and their inhibitors. *Wound Repair Regen* 1999, 7, 442-452.

[95] Diegelmann, RF. Excessive neutrophils characterize chronic pressure ulcers. *Wound Repair Regen*2003, 11, 490-495.

[96] Pattison, DI; Davies, MJ. Reactions of myeloperoxidase-derived oxidants with biological substrates: gaining chemical insight into human inflammatory diseases. *Curr Med Chem* 2006, 13, 3271-3290.

[97] Fu, X; Kassim, SY; Parks, WC; Heinecke, JW. Hypochlorous acid oxygenates the cysteine switch domain of pro-matrilysin (MMP-7). *J Biol Chem* 2001, 276, 41279-41287.

[98] Hirche, TO; Gaut, JP; Heinecke, JW; Belaaouaj, A. Myeloperoxidase plays xritical roles in killing *Klebsiella pneumoniae* and inactivating neutrophil elastase: effects on host defense. *J Immunol* 2005, 174, 1557-1565.

[99] Hirche, TO; Benabid, R; Deslee, G; Gangloff, S; Achilefu, S; Guenounou, M; Lebargy, F; Hancock, RE; Belaaouaj, A. Neutrophil elastase mediates innate host protection against P*seudomonas aeruginosa. J Immunol* 2008, 181, 4945-4954.

[100] Lincoln, JA; Lefkowitz, DL; Cain, T; Castro, A; Mills, KC; Lefkowitz, SS; Moguilevsky, N; Bollen, A. Exogenous myeloperoxidase enhances bacterial phagocytosis and intracellular killing by macrophages. *Infect Immun* 1995, 63, 3042-3047.

[101] Lefkowitz, SS; Gelderman, MP; Lefkowitz, DL; Moguilevsky, N; Bollen, A. Phagocytosis and intracellular killing of *Candida albicans* by macrophages exposed to myeloperoxidase. *J Infec Dis* 1996, 173, 1202-1207.

[102] Lefkowitz, DL; Gelderman, MP; Fuhrmann, SR; Graham, S; Starnes Iii, JD; Lefkowitz, SS; Bollen, A; Moguilevsky, N. Neutrophilic myelo peroxidase–macrophage interactions perpetuate chronic inflammation associated with experimental arthritis. *Clin Immunol* 1999, 91, 145-155.

[103] Grattendick, K; Stuart, R; Roberts, E; Lincoln, J; Lefkowitz, SS; Bollen, A; Moguilevsky, N; Friedman, H; Lefkowitz, DL. Alveolar macrophage activation by myeloperoxidase. *Am J Resp Cell Mol* 2002, 26, 716-722.

[104] Lau, D; Mollnau, H; Eiserich, JP; Freeman, BA; Daiber, A; Gehling, UM; Brümmer, J; Rudolph, V; Münzel, T; Heitzer, T; Meinertz, T; Baldus, S. Myeloperoxidase mediates neutrophil activation by association with CD11b/CD18 integrins. *PNAS* 2005, 102, 431-436.

[105] El Kebir, D; József, L; Pan, W; Filep, JG. Myeloperoxidase delays neutrophil apoptosis through CD11b/CD18 integrins and prolongs inflammation. *Circ Res* 2008, 103, 352-359.

[106] Klinke, A; Nussbaum, C; Kubala, L; Friedrichs, K; Rudolph, TK; Rudolph, V; Paust, H-J; Schröder, C; Benten, D; Lau, D; Szocs, K; Furtmüller, PG; Heeringa, P; Sydow, K; Duchstein, H-J; Ehmke, H; Schumacher, U; Meinertz, T; Sperandio, M; Baldus, S. Myeloperoxidase attracts neutrophils by physical forces. *Blood* 2011, 117,1350-1358.

[107] Williams, MR; Azcutia, V; Newton, G; Alcaide, P; Luscinskas, FW. Emerging mechanisms of neutrophil recruitment across endothelium. *Trends Immunol* 2011, 32, 461-469.

[108] Yang, JJ; Preston, GA; Pendergraft, WF; Segelmark, M; Heeringa, P; Hogan, SL; Jennette, JC; Falk, RJ. Internalization of proteinase 3 is concomitant with endothelial cell apoptosis and internalization of myeloperoxidase with generation of intracellular oxidants. *Am J Pathol* 2001, 158,581-592.

[109] Astern, JM; Pendergraft I, WF; Falk, RJ; Jennette, JC; Schmaier, AH; Mahdi, F; Preston, GA. Myeloperoxidase interacts with endothelial cell-surface cytokeratin 1 and modulates bradykinin production by the plasma Kallikrein-Kinin system. *Am J Pathol* 2007, 171, 349-360.

[110] Haegens, A; Vernooy, JHJ; Heeringa, P; Mossman, BT; Wouters, EFM. Myeloperoxidase modulates lung epithelial responses to pro-inflammatory agents. *Eu Resp J* 2008, 31,252-260.

[111] Wei, Y; Liu, X-m; Peyton, KJ; Wang, H; Johnson, FK; Johnson, RA; Durante, W. Hypochlorous acid-induced heme oxygenase-1 gene expression promotes human endothelial cell survival. *Am J Physiol - Cell Physiol* 2009, 297, C907-C915.

[112] Wu, AHB. Novel biomarkers of cardiovascular disease: myelo peroxidase for acute and/or chronic heart failure? *Clin Chem* 2009, 55, 12-14.

[113] Nicholls, SJ; Hazen, SL. Myeloperoxidase and cardiovascular disease. *Arterioscl Throm Vas* 2005, 25, 1102-1111.

[114] Le Marchand, L; Seifried, A; Lum, A; Wilkens, LR. Association of the myeloperoxidase −463G→A polymorphism with lung cancer risk. *Cancer Epidem Biomar* 2000, 9, 181-184.

[115] Schabath, MB; Spitz, MR; Zhang, X; Delclos, GL; Wu, X. Genetic variants of myeloperoxidase and lung cancer risk. *Carcinogenesis* 2000, 21, 1163-1166.

[116] Madhusudhana Rao, A; Anand, U; Anand, CV. Myeloperoxidase in chronic kidney disease. *Indian J Clin Biochem* 2011, 26, 28-31.

[117] Johnson, KJ; Fantone, JC; Kaplan, J; Ward, PA. In vivo damage of rat lungs by oxygen metabolites. *J Clin Invest* 1981, 67, 983-993.

[118] Sugiyama, S; Okada, Y; Sukhova, GK; Virmani, R; Heinecke, JW; Libby, P. Macrophage myeloperoxidase regulation by granulocyte macrophage colony-stimulating factor in human atherosclerosis and implications in acute coronary syndromes. *Am JPathol* 2001, 158, 879-891.

[119] Nagra, RM; Becher, B; Tourtellotte, WW; Antel, JP; Gold, D; Paladino, T; Smith, RA; Nelson, JR; Reynolds, WF. Immunohistochemical and genetic evidence of myeloperoxidase involvement in multiple sclerosis. *J Neuroimmunol* 1997, 78, 97-107.

[120] Hoy, A; Leininger-Muller, B; Poirier, O; Siest, G; Gautier, M; Elbaz, A; Amarenco, P; Visvikis, S. Myeloperoxidase polymorphisms in brain infarction. Association with infarct size and functional outcome. *Atherosclerosis* 2003, 167, 223-230.

[121] Wu, D-C; Teismann, P; Tieu, K; Vila, M; Jackson-Lewis, V; Ischiropoulos, H; Przedborski, S. NADPH oxidase mediates oxidative stress in the 1-methyl-4-phenyl-1,2,3,6-tetrahydropyridine model of Parkinson's disease. *PNAS* 2003, 100, 6145-6150.

[122] Baldus, S; Heeschen, C; Meinertz, T; Zeiher, AM; Eiserich, JP; Münzel, T; Simoons, ML; Hamm, CW. Investigators, obotC, myeloperoxidase serum levels predict risk in patients with acute coronary syndromes. *Circulation* 2003, 108, 1440-1445.

[123] Brennan, M-L; Penn, MS; Van Lente, F; Nambi, V; Shishehbor, MH; Aviles, RJ; Goormastic, M; Pepoy, ML; McErlean, ES; Topol, EJ; Nissen, SE; Hazen, SL. Prognostic value of myeloperoxidase in patients with chest pain. *New Engl J Med* 2003, 349, 1595-1604.

[124] Tang, WHW; Tong, W; Troughton, RW; Martin, MG; Shrestha, K; Borowski, A; Jasper, S; Hazen, SL; Klein, AL. Prognostic value and echocardiographic determinants of plasma myeloperoxidase levels in chronic heart failure. *J Am Coll Cardiol* 2007, 49, 2364-2370.

[125] Mocatta, TJ; Pilbrow, AP; Cameron, VA; Senthilmohan, R; Frampton, CM; Richards, AM; Winterbourn, CC. Plasma concentrations of myeloperoxidase predict mortality after myocardial infarction. *J Am Coll Cardiol* 2007, 49, 1993-2000.

[126] Heslop, CL; Frohlich, JJ; Hill, JS. Myeloperoxidase and C-reactive protein have combined utility for long-term prediction of cardiovascular mortality after coronary angiography. *J Am Coll Cardiol* 2010, 55, 1102-1109.

[127] Sinning, C; Schnabel, R; Peacock, WF; Blankenberg, S. Up-and-coming markers: myeloperoxidase, a novel biomarker test for heart failure and acute coronary syndrome application? *Congest Heart Fail* 2008, 14, 46-48.

[128] Rudolph, V; Goldmann, BU; Bös, C; Rudolph, TK; Klinke, A; Friedrichs, K; Lau, D; Wegscheider, K; Haddad, M; Meinertz, T; Baldus, S. Diagnostic value of MPO plasma levels in patients admitted for suspected myocardial infarction. *Int J Cardiol* 2011, 153, 267-271.

[129] Lewandrowski, K; Chen, A; Januzzi, J. Cardiac markers for myocardial infarction. A brief review. *Am J Clin Pathol* 2002, 118, S93-S99.

[130] Meuwese, MC; Stroes, ESG; Hazen, SL; van Miert, JN; Kuivenhoven, JA; Schaub, RG; Wareham, NJ; Luben, R; Kastelein, JJP; Khaw, K-T; Boekholdt, SM. Serum myeloperoxidase levels are associated with the future risk of coronary artery disease in apparently healthy individuals: The EPIC-Norfolk prospective population study. *J Am Coll Cardiol* 2007, 50, 159-165.

[131] Koníková, E; Glasová, M; Kusenda, J; Babusíková, O. Intracellular markers in acute myeloid leukemia diagnosis. *Neoplasma* 1998, 45, 82-91.

[132] Leong AS-Y; Cooper KL; W-M, F. *Manual of diagnostic antibodies for immunohistology*. Greenwich Medical Media: London; 2003, 325-326.

[133] Suzuki, K; Ota, H; Sasagawa, S; Sakatani, T; Fujikura, T. Assay method for myeloperoxidase in human polymorphonuclear leukocytes. *Anal Biochem* 1983, 132, 345-352.

[134] Graff, G; Gamache, DA; Brady, MT; Spellman, JM; Yanni, JM. Improved myeloperoxidase assay for quantitation of neutrophil influx in a rat model of endotoxin-induced uveitis. *J Pharmacol Toxicol* 1998, 39, 169-178.

[135] Babu, S; Mohapatra, S; Zubkov, L; Murthy, S; Papazoglou, E, A PMMA microcapillary quantum dot linked immunosorbent assay (QLISA). *Biosens Bioelectron* 2009, 24, 3467-3474.

[136] Lin, K-C; Kunduru, V; Bothara, M; Rege, K; Prasad, S; Ramakrishna, BL. Biogenic nanoporous silica-based sensor for enhanced electrochemical detection of cardiovascular biomarkers proteins. *Biosens Bioelectro* 2010, 25, 2336-2342.

[137] McDonnell, B; Hearty, S; Finlay, WJJ; O'Kennedy, R. A high-affinity recombinant antibody permits rapid and sensitive direct detection of myeloperoxidase. *Anal Biochem* 2011, 410, 1-6.

[138] Liu, B; Lu, L; Li, Q; Xie, G. Disposable electrochemical immunosensor for myeloperoxidase based on the indium tin oxide electrode modified with an ionic liquid composite film containing gold nanoparticles, poly(o-phenylenediamine) and carbon nanotubes. *Microchim Acta* 2011, 173, 513-520.

[139] Posthuma-Trumpie, G; Korf, J; van Amerongen, A. Lateral flow (immuno)assay: its strengths, weaknesses, opportunities and threats. A literature survey. *Anal Bioanal Chem* 2009, 393, 569-582.

[140] Venkatraman, VL; Reddy, RK; Zhang, F; Evans, D; Ulrich, B; Prasad, S. Iridium oxide nanomonitors: Clinical diagnostic devices for health monitoring systems. *Biosens Bioelectron* 2009, 24, 3078-3083.

[141] Windmiller, JR; Chinnapareddy, S; Santhosh, P; Halámek, J; Chuang, M-C; Bocharova, V; Tseng, T-F; Chou, T-Y; Katz, E; Wang, J. Strip-based amperometric detection of myeloperoxidase. *Biosens Bioelectro* 2010, 26, 886-889.

[142] Weber, SA; Watermann, N; Jossinet, J; Byrne, JA; Chantrey, J; Alam, S; So, K; Bush, J; O'Kane, S; McAdams, ET. Remote wound monitoring of chronic ulcers. *IEEE Trans Inf Technol Biomed* 2010, 14, 371-377.

[143] Hasmann, A; Gewessler, U; Hulla, E; Schneider, KP; Binder, B; Francesko, A; Tzanov, T; Schintler, M; Van der Palen, J; Guebitz, GM; Wehrschuetz-Sigl, E. Sensor materials for the detection of human neutrophil elastase and cathepsin G activity in wound fluid. *Exp Dermatol* 2011, 20, 508-513.

[144] Garcia, S; McNulty, A; Villarreal, K. Devices and methods for the diagnosis and treatment of wounds using biomarkers. *KCI Licensing, Inc* 2011, United States US 2012/0129786 A1.

[145] Gübitz, G; Wehrschütz-Sigl, E; Hasmann, A; Binder, B; Schintler, M; Burnett, M. Method for detecting a wound infection using substrates for at least two enzymes. *Sonn & Patentanwälte* 2010, Austria PCT/EP2010/056811.

[146] Zhang, C; Yang, J; Jacobs, JD; Jennings, LK. Interaction of myeloperoxidase with vascular NAD(P)H oxidase-derived reactive oxygen species in vasculature: implications for vascular diseases. *Am J Physiol - Heart C* 2003, 285, H2563-H2572.

[147] Li, Y; Trush, MA. Diphenyleneiodonium, an NAD(P)H oxidase inhibitor, also potently inhibits mitochondrial reactive oxygen species production. *Biochem Bioph Res Co* 1998, 253, 295-299.

[148] Hori, H; Fenna, RE; Kimura, S; Ikeda-Saito, M. Aromatic substrate molecules bind at the distal heme pocket of myeloperoxidase. *J Biol. Chem.* 1994, 269, 8388-8392.

[149] Segelmark, M; Persson, B; Hellmark, T; Wieslander, J. Binding and inhibition of myeloperoxidase (MPO): a major function of ceruloplasmin? *Clin Exp Immunol* 1997, 108, 167-174.

[150] Daphna, EM; Michaela, S; Eynat, P; Irit, A; Rimon, S. Association of myeloperoxidase with heparin: oxidative inactivation of proteins on the surface of endothelial cells by the bound enzyme. *Mol Cell Biochem* 1998, 183, 55-61.

[151] Tidén, A-K; Sjögren, T; Svensson, M; Bernlind, A; Senthilmohan, R; Auchère, F; Norman, H; Markgren, P-O; Gustavsson, S; Schmidt, S; Lundquist, S; Forbes, LV; Magon, NJ; Paton, LN; Jameson, GNL; Eriksson, H; Kettle, AJ. 2-Thioxanthines are mechanism-based inactivators of myeloperoxidase that block oxidative stress during inflammation. *J Biol Chem* 2011, 286, 37578–37589.

[152] Burner, U; Obinger, C; Paumann, M; Furtmüller, PG; Kettle, AJ. Transient and steady-state kinetics of the oxidation of substituted benzoic acid hydrazides by myeloperoxidase. *J Biol Chem* 1999, 274, 9494-9502.

[153] Kettle, AJ; Gedye, CA; Winterbourn, CC. Mechanism of inactivation of myeloperoxidase by 4-aminobenzoic acid hydrazide. *Biochem J* 1997, 321, 503-508.

[154] Schaffer, S; Eckert, GP; Müller, WE; Llorach, R; Rivera, D; Grande, S; Galli, C; Visioli, F. Hypochlorous acid scavenging properties of local Mediterranean plant foods. *Lipids* 2004, 12, 1239-1247.

[155] Van Antwerpen, P; Boudjeltia, KZ; Babar, S; Legssyer, I; Moreau, P; Moguilevsky, N; Vanhaeverbeek, M; Ducobu, J; Nève, J. Thiol-containing molecules interact with the myeloperoxidase/H_2O_2/chloride system to inhibit LDL oxidation. *Biochem Bioph Res Com* 2005, 337, 82-88.

[156] Francesko, A; Soares da Costa, D; Lisboa, P; Reis, RL; Pashkuleva, I; Tzanov, T. GAGs-thiolated chitosan assemblies for chronic wounds treatment: control of enzyme activity and cell attachment. *J Mater Chem* 2012, 22, 19438-19446.

[157] Meotti, FC; Senthilmohan, R; Harwood, DT; Missau, FC; Pizzolatti, MG; Kettle, AJ. Myricitrin as a substrate and inhibitor of myeloperoxidase: Implications for the pharmacological effects of flavonoids. *Free Radical Bio Med* 2008, 44, 109-120.

[158] Rice-Evans, CA; Miller, NJ; Paganga, G. Structure-antioxidant activity relationships of flavonoids and phenolic acids. *Free Radical Bio Med* 1996, 20, 933-956.

[159] Mira, L; Fernandez, MT; Santos, M; Rocha, R; Florêncio, MH; Jennings, KR. Interactions of flavonoids with iron and copper ions: a mechanism for their antioxidant activity. *Free Radical Res* 2002, 36, 1199-1208.

[160] Pauff, JM; Hille, R. Inhibition studies of bovine xanthine oxidase by luteolin, silibinin, quercetin, and curcumin. *J Nat Prod* 2009, 72, 725-731.

[161] Kato, Y; Nagao, A; Terao, J; Osawa, T. Inhibition of myeloperoxidase-catalyzed tyrosylation by phenolic antioxidants *in vitro*. *Biosci Biotech Bioch* 2003, 67, 1136-1139.

[162] Haslam, E. Natural polyphenols (vegetable tannins) as drugs: possible modes of action. *J Nat Prod* 1996, 59, 205-215.

[163] Díaz-GonzáLez, M; Rocasalbas, G; Francesko, A; Touriño, S; Torres, JL; Tzanov, T. Inhibition of deleterious chronic wound enzymes with plant polyphenols. *Biocatal Biotransform* 2012, 30, 102-110.

[164] Francesko, A; Rocasalbas, G; Touriño, S; Mattu, C; Gentile, P; Chiono, V; Giardelli C; Tzanov, T. Cross-linked collagen sponges loaded with plant polyphenols with inhibitory activity towards chronic wound enzymes. *Biotechnol J* 2011, 6, 1208-1218.

[165] Francesko, A; Soares da Costa, D; Reis, R; Pashkuleva, I; Tzanov, T. Functional biopolymer-based matrices for modulation of chronic wound enzyme activities. *Acta Biomater* 2012, accepted for publication.

[166] Kettle, AJ; Winterbourn, CC. Mechanism of inhibition of myeloperoxidase by anti-inflammatory drugs. *Biochem Pharmacol* 1991, 41, 1485–1492.

[167] Nève, J; Parij, N; Moguilevsky, N. Inhibition of the myeloperoxidase chlorinating activity by non-steroidal anti-inflammatory drugs investigated with a human recombinant enzyme. *Eur J Pharmacol* 2001, 417. 37–43.

[168] Sliskovic, I; Abdulhamid, I; Sharma, M; Abu-Soud, HM. Analysis of the mechanism by which tryptophan analogs inhibit human myeloperoxidase. *Free Radical Bio Med* 2009, 47, 1005–1013.

[169] Rees, MD; Bottle, SE; Fairfull-Smith, KE; Malle, E; Whitelock, JM; Davies, MJ. Inhibition of myeloperoxidase-mediated hypochlorous acid production by nitroxides. *Biochem. J* 2009, 421, 79–86.

[170] Koelsch, M; Mallak, R; Graham, GG; Kajer, T; Milligan, MK; Nguyen, LQ; Newsham, DW; Keh ,JS; Kettle, AJ; KScott, KF; Ziegler, JB; Pattison, DI; Fu, S; Hawkins, CL; Rees, MD; Davies, MJ. Acetaminophen (paracetamol) inhibits myeloperoxidase-catalyzed oxidant production and biological damage at therapeutically achievable concentrations. *Biochem Pharmacol* 2010, 79, 1156–1164.

[171] Kagan, VE; Konduru, NV; Feng, W; Allen, BL; Conroy, J; Volkov, Y; Vlasova, II; Belikova, NA; Yanamala, N; Kapralov, A; Tyurina, YY; Shi, J; Kisin, ER; Murray, AR; Franks, J; Stolz, D; Gou, P; Klein-Seetharaman, J; Fadeel, B; Star, A; Shvedova, AA. Carbon nanotubes degraded by neutrophil myeloperoxidase induce less pulmonary inflammation. *Nat Nanotechnol* 2010, 5, 354-359.

[172] Vlasova, II; Vakhrusheva, TV; Sokolov, AV; Kostevich, VA; Ragimov, AA. Peroxidase-induced degradation of single-walled carbon nanotubes: hypochlorite is a major oxidant capable of in vivo. *J Physics: Conference Series* 2011, 291, 012056.

[173] Shvedova, AA; Kapralov, AA; Feng, WH; Kisin, ER; Murray, AR; Mercer, RR; St Croix, CM; Lang, MA; Watkins, SC; Konduru, NV; Allen, BL; Conroy, J; Kotchey, GP; Mohamed, BM; Meade, AD; Volkov, Y; Star, A; Fadeel, B; Kagan, VE. Impaired clearance and enhanced pulmonary inflammatory/fibrotic response to carbon nanotubes in myeloperoxidase-deficient mice. *Plos one* 2012, 7, e30923.

In: Enzymes and Enzyme Activity
Editor: Eva M. Lashinski

ISBN: 978-1-62417-696-8

Chapter II

Monitoring the Activation of Calpain I for Sensitive Assessment of Pathogenesis and Neuroprotectant Action

Ben A. Bahr*
William C. Friday Laboratory,
Biotechnology Research and Training Center
University of North Carolina – Pembroke,
Pembroke, North Carolina US

Abstract

Calpain is a calcium-sensitive protease whose over-activation has been linked to a variety of degenerative conditions. This chapter will focus on calpain's involvement in neurodegeneration, however its pathogenic role is not restricted to the nervous system. It has been implicated in cardiac pathology, kidney hypoxia, respiratory disorders, myopathy, bone erosion, and eye diseases. Of the many substrates that calpain cleaves in brain and peripheral tissues, the cytoskeletal protein

* Corresponding author: B. A. Bahr, Biotechnology Research and Training Center, University of North Carolina - Pembroke, Pembroke, NC 28372-1510; email: Bahr@uncp.edu.

αII-spectrin has been identified as being useful to measure pathogenic calpain activation during cellular toxicity and certain diseases. Calpain-mediated spectrin breakdown has been used extensively to monitor cytoskeletal damage, one of the sensitive precursors of neuronal pathology found in animal models of excitotoxic insults and human brain injury. Cleavage of spectrin by calpain generates a distinct set of large breakdown products (BDPs) that are unusually stable. Detection and measurements of BDPs have allowed calpain-mediated proteolysis to be established as an early event contributing to the neurodegeneration associated with ischemia, hypoxia, genetic defects, vasospasm, traumatic brain injury, and seizures, and it may be involved in brain aging and Alzheimer-type pathogenesis. Being sensitive markers of excitotoxicity and other types of cellular deterioration, the spectrin fragments make up an important assay of pathology using specific labeling by antibodies against the cleavage site recognized by calpain. The ability to detect the earliest signs of pathology in experimentally-induced models of disease states allows for the careful screening of potential therapeutic agents. Especially in brain tissue targeted by acute, definable neuronal insults, neuroprotectants can be assessed for their ability to reduce calpain-mediated cytoskeletal damage and associated damage to central synapses. Excitotoxic calpain activation is commonly associated with reductions in synaptic markers, particularly among pyramidal neurons of the hippocampus. In fact, calpain-mediated spectrin breakdown levels have been found to correspond with synaptic deterioration as well as with subsequent cell death in the hippocampal CA1 subfield. The excitotoxic hippocampal slice model is often utilized for neuroprotectant characterization and screening, in which slice cultures are briefly exposed to an excitotoxin, and the neuronal stimulation subsequently quenched with glutamate receptor antagonists. Potential neuroprotectants, applied to the slices before or after the insult, are tested for effects on cytoskeletal proteolysis and synaptic decline evident 2-48 hours later. Monitoring spectrin breakdown and synaptic markers in *in vitro* and *in vivo* disease models is a valuable first step in identifying lead compounds for therapeutic indications. They are also sensitive measures for preclinical toxicology testing.

Introduction

Over-activation of calpain I, a calcium-activated proteolytic enzyme, has been linked to many types of degenerative insults in different organs. Calpain I, also known as μ-calpain, is found in dendrites and undergoes autolytic activation (Suzuki *et al.*, 1981; Inomata *et al.*, 1986, 1988; Perlmutter *et al.*,

1990; Saido *et al.*, 1992). Such activation occurs at micromolar calcium concentrations that occur in the postsynaptic environment during the stimulation of glutamatergic synapses (Guthrie *et al.*, 1991). Calpain appears to have access to multiple substrates and has been implicated in signal transduction events related to cell structure (Fox *et al.*, 1993; Inomata *et al.*, 1996; Kawasaki *et al.*, 1997; Bahr, 2000; Wang and Yuen, 1999). Dozens of substrates for calpain have been identified and several of them have been used to determine the role of the protease in experimentally-induced and naturally-occurring pathologies. Besides the growing list of neurological conditions linked to calpain (see Vanderklish and Bahr, 2000; Liu *et al.,* 2008), a board array of studies also suggests that calpain activation is involved in i) spinal cord injury (Li *et al.*, 1996a; Banik *et al.*, 1997), ii) cardiac pathologies (DiLisa *et al.,* 1995; Atsma *et al.,* 1995; Arthur and Belcastro, 1997), iii) kidney damage (Edelstein *et al.*, 1996, 1997a; Kelton *et al.*, 1996), iv) respiratory disorders (Miller *et al.,* 1995; Jiang *et al.*, 1998), v) myopathies (Kumamoto *et al.,* 1995; Beckmann *et al.*, 1996; Kumamoto *et al.*, 1997), vi) transplant complications (Kohli *et al.*, 1997; Tijsen *et al.*, 1997), vii) bone erosion (Amar *et al.*, 1996), viii) glaucoma (Govindarajan *et al.*, 2008), ix) cataract formation (Wang *et al.*, 1996), and x) apoptotic events (Martin *et al.,* 1995; Kluck *et al.*, 1997).

A primary trigger of pathology in most paradigms in the field is a rise in the concentration of intracellular calcium this has been used to study the responses of neuronal populations to acute, definable insults (e.g. ischemia; see Siesjo, 1988). Aged humans and age-related pathologies, including Alzheimer's disease, have been linked to disturbances in calcium homeostasis within their tissues. Leupeptin-sensitive proteolysis of a large protein doublet in brain synaptic membranes can be stimulated with low levels of calcium (Baudry et al., 1981). Later, the doublet was identified as spectrin, which is a major actin-binding cytoskeletal component of the neuronal membrane, and a favored substrate for calpain (Siman *et al.*, 1984; Biovin *et al.*, 1990), particularly when calmodulin is present (Seubert *et al.*,1987). The αII-subunit is cleaved by calpain between Tyr^{115} and Gly^{116} (Harris *et al.*, 1989) of the 11th spectrin repeat unit (see Moon and McMahon, 1990) resulting in two stable breakdown products (BDPs or SBDPs) of nearly equal electrophoretic mobility. Besides spectrin cleavage, calpain has been implicated in fragmenting neurotransmitter receptor clustering molecules and synaptic adhesion molecules (Capaldi *et al.,* 1997; Hoffman *et al.*, 2001; Ryzhikov and Bahr, 2008). Cytoskeletal and synaptic protein cleavage has been linked to a vast array of pathogenic events.

Research has shown that polyclonal antibodies developed and affinity-purified against native spectrin recognize the calpain-mediated fragments on western blots, and quantification of the immunolabeled cleavage products has been used in depth to measure calpain activation *in situ*. Breakdown product-specific antibodies have also been developed to epitopes exposed by the calpain cleavage event (Saido *et al.*, 1993; Roberts-Lewis *et al.*, 1994; Bahr *et al.*, 1994c, 1995). Antibodies to the five amino acid sequence (QQEVY) at the carboxyl-end of the amino-terminal fragment (anti-BDP_N) distinguish an antigen of 150 kDa in brain membrane samples treated with calcium to produce endogenous calpain. Comparably, antibodies to the calpain-rendered epitope (residues GMMPR) of the carboxy-terminal fragment (anti-BDP_C) label a 155 kDa antigen in calcium-treated, but not control, membranes. Thus, antibodies against new spectrin termini exposed by calpain proteolysis selectively label distinct proteolytic fragments. These fragments correspond to those described all through the literature as BDP1 and BDP2 (Seubert *et al.*, 1988ab, 1989ab; Siman and Noszek, 1988). Occasionally, antibodies to native spectrin also stain a third lower molecular weight calcium-induced cleavage product, creating a triplet pattern with the larger two fragments (Roberts-Lewis *et al.*, 1994; Bahr *et al.*, 1995). Adjacent immunoblot lanes have been stained with anti-spectrin vs. anti-BDP antibodies to indicate that the middle band (150 kDa) of the triplet corresponds with the BDP_N antigen (Bahr *et al.*, 1995; Bahr, 2000).

Calpain overactivation is a critical factor in neuropathology. The pathogenic relevance depends largely on such factors as activation threshold, subcellular distribution, and the extent to which the proteolytic consequences are reversible.

A common observation linked to the types of neuropathology noted above is the calcium-dependent proteolysis of spectrin by calpain, with the resultant production of 150 kDa breakdown products (BDPs). The BDPs are relatively stable and can be freely measured after immunolabeling with antibodies raised to intact spectrin (Siman *et al.*, 1984; Seubert *et al.*, 1988ab, 1989ab; Siman and Noszek, 1988) or to peptide sequences on either side of the calpain cleavage site (Saido *et al.*, 1993; Roberts-Lewis *et al.*, 1994; Bahr *et al.*, 1994c, 1995). Rapid and sustained increases in the levels of spectrin BDPs have been observed across a broad spectrum of experimentally induced and insidious neuropathologies to be reviewed here.

Combined, these studies implicate calpain activation as a primary supplier to the onset and progression of degeneration in vulnerable populations of neurons. Furthermore, they have provided a sensitive and economical technique for detecting the earliest stages of pathogenesis.

Normal vs. Over-Activation of Calpain

Functionally relevant calpain activation can be expected to be a highly localized and selective event *in vivo* for several reasons: i) calcium elevation in neurons is strongly regulated by energy-dependent buffering and extrusion systems, and can be sorted by dendritic structure; ii) routes of calcium entry (e.g. voltage-gated and receptor-operated channels) exhibit distinct distributions, often restricted to particular lamina or neural populations; and iii) inhibitory (calpastatin) and stimulatory factors modulate the activity and most likely the substrate selectivity of calpain (Takeyama et al., 1986; Melloni *et al.*, 1998; Liu et al., 2008). However, given the broad list of calpain substrates and the irreversible nature of proteolysis, a challenge to any one of the above systems could permit more generalized calpain activation, causing inherently pathological rather than functional proteolysis. The activation threshold of calpain I is well above the resting $[Ca^{2+}]_i$ (~80 nM), making it unlikely that the protease would be activated by fluctuations in calcium concentrations associated with routine synaptic activity, mediated predominantly by AMPA receptors. Indeed, under control conditions BDPs are at low or undetectable levels in immunoblot assays of adult brain homogenate (generally less than 2% of the total anti-spectrin immunoreactivity), a finding upheld by immunohistochemical methods. However, western blot analyses have shown that BDPs are produced in hippocampus during direct (Seubert *et al.*, 1988ab; del Cerro *et al.*, 1994) or indirect (Bahr *et al.*, 1995) stimulation of NMDA receptors.

The potent blocking action of calpain inhibitors on long-term potentiation suggests that calpain may become activated during physiological NMDA receptor stimulation (del Cerro *et al.*, 1990; Denny *et al.*, 1990). The implied relationship between calpain activation and synaptic transmission conditions that evoke large increases in $[Ca^{2+}]_i$ (i.e. high frequency activity) was established *in situ* using BDP-specific antibodies (Vanderklish *et al.*, 1995). Spectrin BDPs were found to accumulate in postsynaptic elements of CA1 pyramidal cells to a range that correlated with the number of short, high frequency bursts of afferent stimulation delivered to Schaffer-collateral projections; low frequency stimulation did not activate calpain. Higher resolution studies employing a calpain substrate that exhibits fluorescence resonance energy transfer indicate that activation of calpain occurs in dendritic spine heads following stimulation of NMDA receptors (Vanderklish *et al.*, 2000).

The above results emphasize that calpain activation in adult glutamatergic neurons most likely has a physiological role, but is a relatively infrequent event, potentially reserved for specialized synaptic modulation events. They also indicated that spectrin BDPs can be induced by modest levels of NMDA receptor activation and associated calcium influx. Interestingly, in the mutant brindled mouse with altered calcium regulation, spectrin BDPs appear between postnatal days 12 and 15 in brain regions that exhibit degeneration with further maturation (Seubert *et al.*, 1990).

Hippocampal Slice Model of Excitotoxicity

Intense activation of NMDA-type glutamate receptors leads to spectrin proteolysis (Seubert *et al.*, 1988b; del Cerro *et al.*, 1994). This effect was shown in studies utilizing organotypic hippocampal slice cultures (Stoppini *et al.,* 1991; Bahr, 1995) and is correlated with the amount of stimulation (del Cerro *et al.*, 1994; Vanderklish *et al.*, 1995). The spectrin breakdown can be detected after only a few minutes of receptor activation (Bahr *et al.*, 1995; Bahr, 2000; Caba *et al.*, 2002), far in advance of histological signs of degeneration observed following administration of NMDA *in vivo* (Siman and Noszek, 1988; Siman *et al.*, 1989).

It also precedes the onset of propidium iodide staining evident after lengthy exposure to NMDA (Vornov *et al.*, 1991). As expected, both the proteolytic (Seubert *et al.*, 1988b; Siman *et al.,* 1989; del Cerro *et al.*, 1994; Bahr, 1995) and degenerative (Siman *et al.,* 1989) responses to excessive NMDA exposure are blocked by NMDA receptor antagonists.

Excitotoxic exposures in the slice model consistently lead to calpain activation and the characteristic spectrin breakdown events (Bahr et al., 1994a, 1995, 2002; Caba *et al.*, 2002; Karanian *et al.*, 2005, 2006), and the calpain response is often associated with the following:

- loss of pre-synaptic proteins
- loss of postsynaptic receptors
- disruption of cell survival signaling
- enhanced vulnerability to a secondary toxic insult
- enhanced vulnerability to protein accumulation pathology

The rapid calpain activation and BDP induction in response to NMDA raises the question of what degree of NMDA-induced proteolysis can be tolerated by neurons for appreciable recovery. Electrophysiological studies suggest that the temporal limit is less than 15 minutes during exposure to saturating concentrations of NMDA (Blake *et al.*, 1988). In cultured hippocampal slices it was demonstrated that antisense-mediated reduction of calpain I by 50% improved the recovery of synaptic potentials following a 15-minute NMDA infusion (Bednarski *et al.*, 1995). Translational suppression of calpain also reduced the levels of NMDA-induced BDPs. The neuroprotection provided by calpain suppression indicates that calcium-dependent proteolysis induced at a large number of sites can become pathogenic within a short time. Upon NMDA receptor stimulation *in vitro* and *in vivo*, levels of BDP formation and neurodegeneration correlate with the amount of NMDA administered (Siman *et al.*, 1989; Bahr *et al.*, 1995). More important is the overlap among areas of unusual NMDA vulnerability and regions of intense spectrin BDP induction. Field CA1 exhibits the following: i) the greatest vulnerability to NMDA (Vornov *et al.*, 1995), *ii*) the highest concentration of NMDA receptors in hippocampus, and *iii*) the most robust BDP staining following a short exposure to the excitotoxin. The results dispute that the relative susceptibility of neurons to NMDA and other excitotoxins is defined by the extent to which calpain is activated. The extent of calpain-mediated spectrin breakdown correlates significantly with measures of synaptic pathology and excitotoxic degeneration, especially in the hippocampal CA1 subfield of neurons (Bahr *et al.*, 2002; Karanian *et al.*, 2005, 2006; Naidoo *et al.*, 2011).

Models of Excitotoxin Exposure

The excitotoxin kainic acid (KA) elicits a reproducible, spatially restricted proteolytic response. Large accumulations of BDP have been found in the hippocampus and piriform cortex within 3-16 h of KA injection, continuing for as long as 9 days (Siman *et al.*, 1989; Siman and Noszec, 1988). In the hippocampus, the calpain-mediated proteolysis was greatest at 24 h, with the highest levels of BDP taking place in the CA3 region, and lesser amounts in CA2/CA1 regions, the dentate, and CA1/subicular regions. This pattern matches the profile of cell loss observed in hippocampus a few days after the insult (Nadler *et al.*, 1978; Siman *et al.*, 1988), and agrees with the distribution of high affinity KA receptors. It is important to note the magnitude of spectrin proteolysis and the extent of cell damage co-vary with the dosage of KA, much as observed when NMDA is used as the excitotoxin (Siman *et al.*, 1989); KA, however, is more potent than NMDA as an excitotoxic agent *in vivo*. In conjunction with the temporal and agonist-specific regional correlations between BDP formation and cell death, the findings argue that calpain plays an early and causal role in the development of excitotoxicity. In different rat models of excitotoxin exposure, the extent of calpain-mediated spectrin breakdown strongly connected with both synaptic marker reductions and cognitive decline (see Karanian et al., 2005, 2007).

Calpain-mediated proteolysis has been used to identify conditions that enhance cellular vulnerability, and such conditions appear to have the ability to act synergistically at the point of calpain activation to enhance pathogenesis. For instance, low level nerve agent exposure using soman was found to enhance hippocampal vulnerability to calpain's cleaving of spectrin (Munirathinam and Bahr, 2004). The apparently harmless low-level soman exposure thus leaves the hippocampus sensitive to the types of insults implicated in traumatic brain injury and stroke. In another study focusing on glucocorticoids and neuronal damage, Elliot *et al.* (1993) found that daily injections of corticosterone (CORT) exacerbated KA-induced spectrin proteolysis. A more direct influence on KA-induced calpain activation has been shown for polyamines. Spermine, spermidine, and putrescine were found to increase after KA lesions and other pathological circumstances associated with calpain activation *in vivo* (e.g., ischemia). With the above order of potency, the polyamines stimulated calpain-mediated hydrolysis of endogenous and exogenous protein substrates. This modulation of calpain by polyamines apparently occurs *in vivo*. As reported by Najm *et al.* (1992), injection of an inhibitor (DFMO) of ornithine decarboxylase, the rate-limiting enzyme in the polyamine metabolic pathway, dramatically reduces KA-induced spectrin proteolysis in the hippocampus. DFMO did not completely abolish KA-induced proteolysis, and the results indicate a relationship between neurodegeneration and the early phase of calpain-mediated proteolysis.

While the types of neuronal insults considered above have the common feature of glutamate receptor hyper-stimulation at an early stage, it must be emphasized that routes to pathology which do not directly engage glutamatergic systems also lead to calpain activation. Colchicine, for example, causes neuronal damage by blocking critical transport processes. Intraventricular injections of the drug caused extensive death among granule cells of the dentate gyrus, with much smaller effects in the other hippocampal subdivisions (Steward *et al.*, 1984; Seubert *et al.*, 1989b); these effects developed over several days. Spectrin BDPs appeared within 24 h of intrahippocampal colchicine injection, peaked at 4 days, and stayed elevated for weeks (Seubert *et al.*, 1989b). As was the case with cell death, the increase in proteolysis was largest in the dentate gyrus, particularly in the molecular layer. Note that these results, together with data on excitotoxicity and ischemia, establish a correlation between selective vulnerability and early appearance of the BDPs that includes all of the major hippocampal divisions. The pathogenic action of some environmental toxins also seems to be mediated by calpain. Such toxins include trimethyltin (TMT), an industrial byproduct, as well as the heavy metal mercury (Elliget *et al.*, 1994) and maitotoxin (Zhao *et al.*, 1999). Peripheral injections of TMT into rats have been shown to produce calpain-mediated spectrin proteolysis in the hippocampus within 24 h (Seubert *et al.*, 1988b); the BDP formation steadily increased thereafter for two weeks. A large BDP response was observed in the dentate gyrus relative to field CA1, but a correlation with the regional vulnerability to TMT was not established. Using hippocampal slice cultures, constant exposure to TMT and snake venoms induced the spectrin fragments within 3 h, the shortest time tested (Bahr *et al.,* 1995).

Hypoxic and Ischemia Injury

Calpain has been linked to hypoxic/ischemic injuries to the brain (Lee, 1995; (Takeyama et al., 1986; Wang and Yuen, 1999) and to peripheral organs including liver (Bronk and Gores, 1993), heart (Tolnadi and Korecky, 1986), and kidney (Edelstein *et al.*, 1995, 1996, 1997ab). Much of the research has focused on the brain since the ischemic condition known as stroke affects 500,000 people each year in the United States alone, making it the third leading cause of death and the number one cause of disability. Patterns of regional vulnerability to ischemia in the brain and temporal characteristics of the degenerative response have been described in detail (see Hatakeyama *et al.*, 1988).

In a typical gerbil model of global forebrain ischemia, transient carotid artery occlusion produced degeneration in a set of highly vulnerable neural populations. With increasing durations of arterial occlusion, cortical and subcortical structures of lesser susceptibility were recruited among the more vulnerable regions in a reliable sequence. Damage to these structures occur at a threshold duration of ischemia, beyond which neurons died after a delay of a few days.

Ischemic lesions were induced in the CA1 subfield of hippocampus, a region which, as noted, contains high levels of NMDA receptors. Blockade of NMDA receptors by antagonists has been shown to ameliorate ischemic injury (Kochhar *et al.*, 1991; Madden *et al.*, 1993; Small and Buchan, 1997), and given that ischemia can lead to massive glutamate efflux, the particularly high density of these receptors in CA1 seems to explain the unusual sensitivity of pyramidal cells in this region. Such results linking the NMDA receptor as a variable in ischemic brain damage serve to strengthen the longstanding hypothesis that calcium influx is critically involved in pathogenesis. In a study that linked the NMDA receptor-dependency of ischemia with the known downstream effects of the calcium ion on calpain activation, Seubert *et al.* (1989a) established that episodes of global ischemia lasting 10 minutes elicit a rapid proteolysis of spectrin. Using antibodies to native spectrin, BDPs were detected on western blots of samples collected 15 minutes after reperfusion. Calpain activity was most pronounced in CA1, while areas of lesser vulnerability (e.g., thalamus and cerebellum) had consistently smaller accumulations of the spectrin fragments. At subsequent time points, BDPs were shown to persist only in CA1, and after 4 days, this region experienced a second, more intense wave of proteolysis.

Under conditions where baseline calpain activation that accompanies acute slice preparation was kept to a minimum, BDP concentrations were approximately doubled in hippocampal slices after 5 minutes of hypoxia (Arai *et al.*, 1991; Lee *et al.*, 1991). Calpain-mediated spectrin proteolysis after global ischemia thus appears to be unconnected to reperfusion, but seems a primary response to hypoxia which is bi-phasic in nature: an early component started during or shortly after ischemia, followed by a later and larger wave correlated with delayed cell death. This general profile holds for multiple models of focal and global ischemia. Using similar techniques, Lee and coworkers (Hong *et al.*, 1994a) have shown that focal neocortical ischemia (achieved by occlusion of both carotids and one middle cerebral artery) is associated with a rapid and sustained induction of spectrin BDPs in the infarct zone (also see Lee *et al.*, 1991; Hong et al., 1994b; Yao *et al.,* 1995 and Bartus *et al.*, 1995ab). More recently, Bartus *et al.* (1998) measured BDP formation in CA1 following four vessel occlusion in the rat and correlated this measure with histological indeces of cell damage and *ex vivo* electrophysiological measures of pyramidal cell synaptic function. Post-ischemic calpain activation correlates with the progressive compromise of pyramidal cells.

The above findings have been corroborated and extended by two groups using BDP-specific antibodies (Saido *et al.*, 1993; Roberts-Lewis *et al.*, 1994).

In immunohistochemical assays, these antibodies labeled only those sites where BDPs accumulated, and thus provided greater spatial resolution of calpain activation, both at the regional and cellular level. Exploiting this specificity, the two groups showed that calpain becomes activated in the dendrites of pyramidal cells in the CA2/CA3 border zone after ischemic episodes as short as 1 min. Longer durations, 5 and 10 minutes, elicited staining in CA1, then cortical, thalamic, and other sites (Roberts-Lewis *et al.*, 1994). Thus, sampling of post-ischemic brain throughout an extended period revealed that CA1 experiences an early phase (within 15 min; Saido *et al.*, 1993), then a later and more intense phase of proteolysis between 1 and 2 days after an intermediate ischemic duration. Slightly longer episodes of carotid artery occlusion elicited this response in additional cortical and thalamic areas (Roberts-Lewis *et al.*, 1994).

Multiple pharmacological approaches have been used to test the receptor dependency of the BDP effect and the prediction made with the assay that calpain is a primary contributor to pathogenesis in the ischemic brain. Inhibition of NMDA receptors with MK-801 has been shown to be effective in reducing both phases of proteolysis in the abovementioned global ischemia paradigm when administered in doses that give significant neuroprotection (Seubert *et al.*, 1990a; Roberts-Lewis and Siman, 1993; Roberts-Lewis *et al.*, 1994). The coordinated reductions in cell death and calpain activation by MK-801 emphasize the significance of NMDA receptor-linked calcium entry in ischemia. It is substantial in this regard that NMDA receptor antagonists also reduce toxicity and spectrin breakdown associated with NMDA administration *in vivo* (Siman *et al.*, 1989), and prevent BDP formation elicited with the excitotoxin *in vitro* (Seubert *et al.*, 1988b; del Cerro *et al.*, 1994; Bahr *et al.*, 1995). Therefore, neuroprotection afforded by NMDA receptor blockade is linked with reductions in calpain activity in two paradigms: ischemia and excitotoxicity. The NMDA receptor channel blocker MK801 repressed calpain activation.

Studies with BDP-specific antibodies represent an important set of observations: *i*) they document with great acuity the strict spatiotemporal correlations between calpain-mediated proteolysis and pathology; *ii*) they indicate proteolysis must persist beyond the initial stimulus-linked phase to irreversibly compromise cells (Saido *et al.*, 1993); and *iii*) they add to previous work in the field suggesting that protection of neurons from ischemia can be accomplished by inhibiting calpain after the primary insult. With regard to the latter, BDP assays have been used to determine the efficacy of protease inhibitors in target neural populations. Inhibition of calpain has been shown to lower calpain-mediated proteolysis and protect hippocampal and cortical neurons from ischemia (Lee *et al.*, 1991; Hong *et al.*, 1994; Bartus *et al.*, 1994ab), with evidence that selective calpain inhibitors do not disrupt normal cell signaling pathways (see Caba et al., 2002). This protection extends to the point of preserving synaptic function inasmuch as calpain inhibitors improve the recovery of synaptic responses after hypoxia in acute brain slices (Arai *et al.*, 1990; Arlinghaus *et al.*, 1991). Additionally, hippocampal slices prepared from animals treated with the calpain inhibitor leupeptin prior to ischemia retained the ability to express long-term potentiation (Lee *et al.*, 1991). This result is of great significance therapeutically as long-term potentiation is a synaptic plasticity thought to cause the encoding of certain types of memory, and the phenomenon itself relies heavily on the structural and functional integrity of excitatory synapses. As predicted from early data obtained with BDP assays - most notably those using the BDP-specific antibodies - blockade of the latter, persistent phase of ischemia-induced proteolysis offers protection of neurons from ischemia (Bartus *et al.*, 1994ab; Markgraf *et al.*, 1998; Inserte et al., 2004). The reports characterizing the phasic generation of spectrin BDPs after ischemia established that the second, persistent wave of proteolysis starting within hours of the first insult-coupled event is correlated with and indicative of eventual cell death. This implies a window of opportunity for therapeutically efficacious post-ischemic inhibition of calpain that may extend many hours after the initial insult.

Traumatic Brain and Spinal Injury

Traumatic brain injury, often shortened to TBI in the medical field, is associated with functional as well as behavioral deficits, axonal pathology, and neuronal death in patients and in animal models of TBI. Predominately involving young people, those that survive a TBI often suffer from emotional, cognitive and motor disruptions. As with other neurological insults described in this chapter, calpain activation occurs within minutes to hours after contusive or diffuse brain trauma, suggesting that calpain-mediated spectrin proteolysis is an initiator and/or early indicator of neuronal damage (see Saatman et al., 2006a, 2010; McGinn et al., 2009). In fact, sustained calpain activation produced by the trauma is associated with cell death and axonal damage in multiple models of TBI. Proteolytic fragments of αII-spectrin mediated by calpain (SBDP145) and caspase-3 (SBDP120) were measured in the cerebrospinal fluid of patients with severe TBI, and the results suggest that spectrin breakdown levels can predict injury severity and mortality after TBI (Mondello et al., 2010). Not only do the spectrin fragments provide strong markers of brain damage and outcome after severe TBI, they may be useful complements to clinical assessment. Among the growing number of pharmacological targets recently evaluated in experimental TBI models, calpain inhibitors have received significant attention as a potential avenue to improve cognitive and motor outcome and reduce brain tissue damage. Using brain trauma models developed by McIntosh and colleagues, Posmantur et al. (1997) demonstrated that calpain inhibitors can attenuate TBI-associated spectrin breakdown as well as motor and cognitive deficits (Saatman *et al.*, 1996b). Perhaps pointing to the extent to which proteolytic damage facilitates functional compromise, behavioral protection occurs before spectrin breakdown is diminished by the calpain inhibition (Saatman et al., 2000). Delayed inhibition of calpain, through direct inhibition or indirectly via targets related to intracellular calcium regulation, also protects against the pathogenic cascade of TBI, attenuating axonal pathology, neuronal death, as well as functional and behavioral deficits (see Saatman *et al.*, 2010). Inhibition of cysteine proteases has also been studied for the treatment of acute and chronic spinal cord injury (Ray *et al.*, 2011).

Hemorrhage and Vasospasm

Calpain-mediated proteolytic responses may also be involved in the neurodegeneration associated with brain trauma and related cerebrovascular problems that can lead to ischemic episodes. As noted, spectrin BDPs appear regionally after experimental brain injury (Saatman *et al.*, 1996a; Kampfl *et al.*, 1996; Posmantur *et al.*, 1997) and in the cerebrospinal fluid of humans following subarachnoid and intraventricular hemorrhages (Lee *et al.*, 1993). Whether the latter are generated solely by neuronal or by vascular tissue as well has yet to be determined. However, increased calpain activation and spectrin hydrolysis have been detected in smooth muscle cells of the rabbit basilar artery after vasospasm (Lee *et al.*, 1993; Yamaura et al., 1993). It is thought that the proteolysis of spectrin, calponin, and regulatory kinases in these cells contributes to the morphological alterations of arterial cells after vasospasm, and perpetuates spastic contraction initiated by hemorrhage (Lee, 1995; Yamaura et al., 1993). In particuluar, proteolytic conversion of protein kinase C into a tonically catalytic form by calpain occurs in basilar artery after subarachnoid hemorrhage (Sato *et al.*, 1997), leading to increased smooth muscle contractility. The evidence suggests that calpain-1 remains activated, and its inhibitor calpastatin reduced, in arterial smooth muscle cells for days after subarachnoid hemorrhage, perpetuating contraction (Yamaura *et al.*, 1993; Sato *et al.*, 1997). With regard to this idea, it has been demonstrated that application of calpain inhibitors attenuates the development and progression of vasospasm (Lee *et al.*, 1997). A dual action of calpain inhibitors on pathological processes in neurons and the vasculature that supplies them could account for their efficacy in a model of brain trauma. The above research indicates that measurements of BDP levels in cerebrospinal fluid may be important in detecting both the neuronal and vascular aspects of cerebrovascular accidents. It also suggest that calpain inhibition can be useful in decreasing the risk of hemorrhage-induced brain damage associated with certain surgical procedures.

Aging and Aging Models

Using anti-spectrin antibodies, an early study (Bahr *et al.*, 1991) tested whether the gradual changes of brain aging are associated with calpain-mediated proteolysis. The study showed that, indeed, BDP levels increase with age in the mouse brain in a regionally specific manner. The hindbrain had low concentrations of BDPs across mice of 3 to 30 months of age, whereas a statistically significant relationship with age was exhibited by the spectrin BDPs from the telencephalic tissue, the same region in which age-related deficits in synaptic markers have been localized (Bahr *et al.,* 1992, 1993; Masliah *et al.,* 1993).

More recently, studies using BDP-specific antibodies were conducted in middle-aged mice with similar results (Vicente et al., 2002). The calpain-mediated spectrin breakdown marker exhibited a pronounced difference among brain regions isolated from the 12-month-old mice ($p<0.001$, Kruskal-Wallis statistic = 19.5), whereas no significant difference in basal levels of the amino-terminal BDP_N was evident across the 3-month samples. The olfactory bulb contained particularly low levels of BDP_N in the 12-month samples. The anti-BDP_N antibody specifically recognized the calpain-mediated fragment and did not label the parent spectrin which was later verified to have migrated to the 240-kDa position on the immunoblots. The age-related increase in BDP_N was found in neocortical and hippocampal tissue samples, but not in the olfactory bulb, thalamus, or hindbrain. In neocortex, the eight middle-aged mice contained three-fold more of the calpain-mediated BDP_N than that found in samples from the nine young mice ($p=0.01$, nonparametric test). Hippocampal samples exhibited a two-fold increase in the spectrin fragment between 3 and 12 months of age ($p<0.01$). The increases in BDP_N at middle age were confirmed by normalizing the individual immunoreactivity values to a standard curve ($r=0.99$; $p<0.0001$) generated from blot samples containing increasing amounts of rat brain membranes positive for spectrin fragmentation. The normalized data exhibited the same fold increases between the 3- and 12-month samples as determined by the integrated density measures. In addition, immunocytochemistry with anti-BDP_N staining showed that the spectrin breakdown selectively accumulates in field CA1 of the hippocampus from middle-aged mice. Focal localization of the calpain-mediated fragment was also found in the cortex. The BDP_N antigen was evident as patterned, punctate immunoreactivity along cortical dendrites in the 12-month tissue.

Additional alterations in the balance of cytoskeletal constituents were found among several structural proteins in the brains of 25-month-old mice as compared to young, 3-month mice (Bahr *et al.*, 1994b). In particular, a 63-kDa tau species was significantly increased in the aged mice which is likely related to a hyperphosphorylated tau isoform. On the other hand, levels of ankyrin, talin, spectrin, and actin were decreased. Across additional age groups of mice, the forebrain reduction in ankyrin was evident at earlier ages than that for spectrin. Note that ankyrin is the linker molecule that anchors integral membrane proteins, including ion pumps, to the spectrin-based membrane cytoskeleton. Like spectrin, ankyrin is also a calpain substrate. In a myocardial ischemia–reperfusion study, calcium-dependent calpain activation was found to degrade both ankyrin and spectrin, leading to sarcolemmal fragility as well as dysfunction of the sarcolemmal sodium pump (Garcia-Dorado et al., 2006).

Together, the age-related alterations to the cytoskeleton likely contribute to cellular and functional compromise in the CNS and other organ systems. Pathogenic processes set in motion by acute neurological insults – including the examples discussed in this chapter – may be recruited in a gradual manner in the aging brain. Age-related disturbances in calcium homeostasis are candidate avenues of pathogenesis, linking excitotoxic vulnerability to age-related cytoskeletal changes that can disrupt ATPase ion pumps. Possibly related to this, Alzheimer's disease and other age-related disorders have a higher risk for stroke events that are known to cause excitotoxic damage. A current treatment for Alzheimer's disease is an NMDA receptor antagonist (memantine) that reduces excitotoxic over-activation of the receptor, and the antagonist was found to reduce both calpain-mediated BDP levels and the decline in synaptic markers produced by 3-nitropropionic acid (a mitochondrial toxin to experimentally model age-related neurodegeneration) (Karanian *et al.,* 2006).

Differences in calcium regulation and associated calpain activation indeed have been measured in fibroblasts isolated from young vs. old humans (Peterson *et al.*, 1991). It has been earlier proposed that dysregulation of the calpain system occurs at a faster rate with age in regions of the brain that exhibit calcium-dependent synaptic plasticities. A long growing hypothesis on the contribution of the calpain system to brain aging suggests other cellular abnormalities (e.g., gradual lysosomal dysfunction, buildup of reactive oxygen species - ROS) that occur with age interact with ongoing, normally sub-threshold neuronal "stresses" (compromised calcium homeostasis, excitotoxicity), the consequences of which may act as pathogenic feedback (see Butler and Bahr, 2006). Three potential feedback pathways that contribute to age-related neurodegeneration include:

I. excitotoxic events – aberrant proteolysis – oxidative stress – enhanced excitotoxic vulnerability. The paired synergies in this positive feedback loop involve

- excitotoxic neuronal activity and ROS production
- proteolytic dysregulation and ROS production
- aberrant proteolysis and toxic oligomerization

II. oxidative stress – lysosomal compromise – mitochondrial dysfunction – further ROS production. The paired synergies in this positive feedback loop involve

- ROS production and lysosomal stress
- cathepsin-caspase dysregulation and mitochondrial enzyme release
- mitochondrial stress and ROS production

III. lysosomal overload – autophagic activation – proteolytic/ROS stress – further lysosomal compromise. Similar synergies is this feedback loop involve

- lysosomal stress and ROS production
- autophagic/lysosomal activation and aberrant proteolysis that leads to ROS

Calpain could contribute to several of the synergies noted, including i) performing proteolytic reactions that facilitate ROS production, ii) generating long-lived protein fragments (e.g. spectrin BDP) that add to the age-related lysosomal overload, and iii) becoming calcium-independent during the disruption of calcium homeostasis to contribute more effectively toward neuronal deterioration. It is noteworthy that calpain-mediated BDPs accumulate relatively early during brain aging, suggesting that aberrant activation of calpain initiates perhaps a set of cellular modifications that lead towards the characteristic synaptic loss and cognitive decline with further aging.

To study the role of calpain in the cascade thought to represent age-related protein accumulation pathology, many studies made use of long-term hippocampal slice cultures. The slice cultures, when treated with low-level chloroquine to experimentally induce lysosomal stress, recapitulate key features of protein accumulation and the associated synaptic and cellular pathology. The slice model is valuable for the testing of hypotheses and compounds with a focus on aging and age-related disorders such as Alzheimer's disease (Bahr et al., 1994a; Bahr, 1995; Bendiske et al., 2002; Butler et al., 2005; Ryzhikov and Bahr, 2008). Besides evidence of calpain activation, the hippocampal slice model of protein accumulation pathology also expresses the following signs of a distinct cascade of neurodegeneration:

- intracellular protein accumulation
- accumulation of protein fragments
- microtubule destabilization
- loss of tubulin-binding proteins
- buildup of tubulin breakdown products (e.g., TN30)
- disruption of microtubule-based transport
- loss of pre-synaptic proteins
- loss of postsynaptic receptors

In human studies, post-mortem tissue exhibited higher concentrations of spectrin BDPs in the hippocampus and neocortex of Alzheimer's disease patients as compared to age-matched individuals with no signs of the disease (Masliah *et al.*, 1990). Activity assays for calpain and calpastatin have shown that this increase may reflect a regionally specific increase in the net calpain activity (Nilsson *et al.*, 1990; also see Haug *et al.*, 1996). Histological measures also suggest that spectrin BDPs increase in areas most susceptible to Alzheimer-type synapse loss. Calpain-mediated proteolysis in such areas may involve mechanisms associated with the types of deafferentation evident in Alzheimer's disease. For example, degradation of spectrin was found to be significantly enhanced within the frontal cortex in a rat model of cholinergic degeneration (Fernandez-Shaw *et al.*, 1997). The events underlying the increased calpain activation in Alzheimer's disease may extend to other somatic contexts. For instance, skin fibroblasts collected from Alzheimer's patients have a significantly higher BDP content than those from age-matched or young donors when stimulated with serum and mitochondrial uncoupling agents (Peterson *et al.*, 1991). Note that this difference was observed despite indications of equal calpain and calpastatin content in the specimens. There was also evidence for an age-related increase in fibroblast BDP content in the healthy subjects. The increased BDP levels in both conditions are most likely related to the alterations in calcium regulation described in skin cells by Peterson and colleagues. Lastly, calpain has been suggested to play a role in tau processing that experiences alterations with age (Mercken *et al.*, 1995; Grynspan *et al.*, 1997; Yang *et al.*, 1997; Bahr and Vicente, 1998). Intraneuronal deposition of tau occurs in the form of paired helical filaments that generate neurofibrillary tangles in Alzheimer's disease.

Example of Neuroprotectant Assessment

Monitoring calpain I activation is very useful for the assessment of neuroprotectant effectiveness. In addition to the studies mentioned in this Chapter in regards to selective calpain inhibitors (Arai et al., 1990; Lee et al., 1991; Li *et al.,* 1993, 1996b; Bartus et al., 1994b; Bahr et al., 1995; Saatman et al., 1996b, Banik et al., 1997; Kawasaki et al., 1997; Posmantur et al., 1997; Markgraf et al., 1998; Wang and Yuen, 1999; Caba et al., 2002; Govindarajan et al., 2008), determining the extent of protection against early calpain activation has been valuable for other avenues of therapeutic development. For an example from the cannabinoid field, positive modulators of endocannabinoid signaling have been well studied for protection against excitotoxic insults related to stroke and seizure damage.

Medicinal indications for cannabinoid drugs have expanded markedly in recent years, including potential treatments for chemotherapy complications, pain, Parkinson's disease, Huntington's disease, Alzheimer's disease, multiple sclerosis, amyotrophic lateral sclerosis, glaucoma, traumatic brain injury, cerebral ischemia, and other excitotoxic insults (see review: Karanian and Bahr, 2006).

CNS cannabinoid receptors are joined by a specific transporter and a fatty-acid amide hydrolase (FAAH) to make up the endocannabinoid system. The transporter and FAAH facilitate cannabinergic inactivation and their inhibitors enhance endocannabinoid signaling.

Neuroprotection in the Hippocampal Slice Model

To test whether inhibitors of endocannabinoid transport (AM404) and FAAH (AM374) enhance CB_1 receptor responses, we used hippocampal slice cultures prepared from rats. The combination of the two inhibitors proved effective for enhancing signal transduction pathways linked to cannabinoid receptors that recognize the endocannabinoids anandamide and 2-arachidonylglycerol, and it was best for excitotoxic protection (Karanian et al., 2005). In the excitotoxic hippocampal slice model that was subjected to excitotoxic stimulation of AMPA-type or NMDA-type glutamate receptors, persistent cytoskeletal damage was indicated by the calpain-mediated BDP_N measured 24-h post-insult. Post-insult activation of cannabinoid responses with direct agonists reduced the cytoskeletal damage. In the AMPA-treated slice cultures, the spectrin breakdown measure was reduced 96% by the AM374/AM404 combination. The dramatic cytoskeletal protection mediated by the neuroprotectant combo and direct agonists was also associated with synaptic protection. In fact, the measured BDP levels had a significant, negative correlation with the postsynaptic marker's immunoreactivity levels (Karanian et al., 2005, 2007). That is, more pathogenic production of BDPs was significantly associated with synaptic decline, and moreover, neuro-protective reduction in BDP levels was associated with preservation of synaptic proteins.

Such a protective association between the diminishing of calpain-mediated cytoskeletal damage and recovery of synaptic integrity has been found with other classes of neuroprotectants (Bahr et al., 2002; Karanian et al., 2006).

Positive modulators of the AMPA receptor-MAPK pathway, Ampakines, were also studied for their neuroprotective ability in the hippocampal slice model. The Ampakine CX516 enhances the AMPA receptor-MAPK axis in cultured hippocampal slices as well as *in vivo*. In slice cultures excessively exposed to the glutamatergic agonist AMPA, specific immunostaining of calpain-mediated BDP was found predominately in CA1 neurons and associated dendritic puncta. Hippocampal slices pre-incubated with CX516 and continually exposed to the Ampakine during the AMPA insult exhibited significantly less BDP staining than slices treated without CX516 (Bahr et al., 2002).

As with *in vivo* models of excitotoxicity (Lee et al., 1991; Roberts-Lewis et al., 1994), the slice model indicates a correspondence between calpain-mediated spectrin breakdown and subsequent cell death in the hippocampal CA1 subfield. BDP immunostaining in the slice model was reduced by CX516 in both the stratum pyramidale layer and the synapse-rich dendritic zone of the stratum radiatum.

Neuroprotection In Vivo

The AM374/AM404 combination for endocannabinoid enhancement was also assessed for cytoskeletal protection *in vivo*. The rat model of excitotoxicity consisted of AMPA being injected unilaterally into the dorsal hippocampus. The dorsal half of the hippocampus was dissected a week later, and the ipsilateral tissue specifically exhibited pronounced BDP levels, reductions in pre- and postsynaptic markers, and a selective decrease in CA1 pyramidal neurons (Karanian et al., 2005). The contralateral hippocampus had no evidence of spectrin BDP_N, nor did it express any synaptic decline as found in the ipsilateral tissue. As in the slice model, the excitotoxic cytoskeletal breakdown *in vivo* corresponded closely with synaptic decline. When a cannabinoid agonist or the AM374/AM404 combination was co-injected with AMPA, spectrin BDP_N was reduced. The combination treatment produced complete cytoskeletal protection in most animals examined. In addition to the protective results with regard to cytoskeletal integrity, additional neuroprotectant effects were found corresponding with reduced BDP levels, including the following:

- postsynaptic glutamate receptor decline evident in the excitotoxic rats was reduced
- excitotoxic reduction in the presynaptic marker synapsin II was nearly completely protected
- a second synaptic vesicle marker, synaptophysin, was significantly recovered
- excitotoxic compromise of signaling pathways (ERK, FAK) was abrogated
- neuronal death and pyknotic changes in the excitotoxic rat hippocampus was reduced
- behavioral deficits exhibited by the excitotoxic animals were reduced

The important observation that blocking endocannabinoid inactivation not only produced cytoskeletal and synaptic protection but also maintained basal levels of extracellular signal regulated-kinase (ERK) and focal adhesion kinase (FAK) pathways indicates a potential protection mechanism through endogenous cell survival signaling. The two kinases are involved in synaptic maintenance and they are disrupted in models of excitotoxicity. Basal levels of activated pFAK and pERK were maintained by the AM374/AM404 combination.

In addition to the cytoskeletal, synaptic, signaling, and cellular protection, two behavioral correlates of excitotoxic brain damage were also reduced by AM374/AM404, implicating endocannabinoid modulation in the protection of brain function. The disturbance expressed as perseverative turning in the unilateral excitotoxic rats was prevented when AM374/AM404 was co-administered with the excitotoxin. In addition, the excitotoxic animals exhibited memory impairment in a fear conditioning paradigm. The cognitive deficit was significantly prevented by the dual blockage of endocannabinoid inactivation. Of particular interest is that the behavioral protection is associated with a corresponding preservation of glutamate receptor subunit GluR1, synaptic vesicle markers, and basal levels of active ERK. And as expected, the reduced calpain-mediated BDP_N levels exhibited a significant correlation with improved performance in the fear conditioning task (Karanian et al., 2005, 2007). An antagonist of the cannabinoid receptor correspondingly abolished the effects of AM374/AM404 on cytoskeletal protection, synaptic protection, maintenance of FAK and ERK/MAPK pathways, and behavioral protection. The antagonist, when administered alone, had no effect on these parameters. Together, the results strongly suggest that the neuroprotection by AM374/AM404 is mediated by responses via the cannabinergic receptor.

Conclusion

Over-activation of calpain I is, in many circumstances, an early and contributing factor in the development of neuropathology. Assays developed for the detection of calpain-mediated spectrin BDPs have been of great value in describing the link between calpain proteolysis and pathology. Calpain activation is an early response to a variety of stressful conditions. If sustained, the proteolytic damage can lead to cell death in vulnerable populations of cells. Measuring spectrin BDPs as markers of calpain-mediated proteolysis has identified the cytoskeletal damage as a commonality among different types of pathologies. Selective detection of the calpain-generated fragments provides the basis of testable predictions and therapeutic development with regards to the early steps of neuropathogenesis as well as the early protection required for effective treatment. The cytoskeletal degradation resulting from excitotoxic insults is known to be associated with subsequent cell death.

Thus, a correspondence exists between the vulnerability to spectrin breakdown and cell death, suggesting that once a certain threshold of cytoskeletal damage is surpassed, the likelihood of cellular recovery is reduced. The ability of a neuroprotectant to reduce cytoskeletal breakdown after an excitotoxic event indeed corresponds with its ability to prevent neuronal death. Accordingly, many studies now point to the calpain-mediated BDP measure as a tool to be exploited for the screening and development of therapeutics against neurodegeneration.

Acknowledgments

The author thanks Alyson Bahr for editorial service and Elizabeth Metzger for technical assistance. This work was supported in part by the National Institutes of Health, the Oliver Smithies Grant from the North Carolina Biotechnology Center (Research Triangle Park), the NASA - North Carolina NC Space Grant, and the William C. Friday Chair endowment.

References

Amar MS, Wishahi HF & Zakhary MM (1996) Clinical and biochemical studies of bone destruction in cholesteatoma. *J Laryngol Otol* 110:534-539.

Arai A, Kessler M, Lee K & Lynch G (1990) Calpain inhibitors improve the recovery of synaptic transmission from hypoxia in hippocampal slices. *Brain Res* 523:313-315.

Arai A, Vanderklish P, Kessler P, Lee K & Lynch G (1991) A brief period of hypoxia causes proteolysis of cytoskeletal proteins in hippocampal slices. *Brain Res* 555:276-280.

Arlinghaus L, Mehdi S & Lee KS (1991) Improved posthypoxic recovery with a membrane-permeable calpain inhibitor. *Eur J Pharmacol* 209:123-125.

Arthur GD & Belcastro AN (1997) A calcium stimulated cysteine protease involved in isoproterenol induced cardiac hypertrophy. *Mol Cell Biochem* 176:241-248.

Atsma DE, Bastiaanse EML, Jerzewski A, Van der Valk LJM & Van der Laarse A (1995) Role of calcium-activated neutral protease (calpain) in cell death in cultured neonatal rat cardiomyocytes during metabolic inhibition. *Circ Res* 76:1071-1078.

Bahr BA (1995) Long-term hippocampal slices: a model system for investigating synaptic mechanisms and pathologic processes. *J Neurosci Res* 42:294-305.

Bahr BA (2000) Integrin-type signaling has a distinct influence on NMDA-induced cytoskeletal disassembly. *J Neuroscience Res* 59:827-832.

Bahr BA & Vicente JS (1998) Age-related phosphorylation and fragmentation events influence the distribution profiles of distinct tau isoforms in mouse brain. *J Neuropathol Exp Neurol* 57:111-121.

Bahr BA, Vanderklish PW, Ha LT, Tin M, Kessler M & Lynch G (1991) Spectrin breakdown products increase with age in telencephalon of mouse brain. *Neurosci Lett* 131:237-240.

Bahr BA, Godshall AC, Hall RA & Lynch G (1992) Mouse telencephalon exhibits an age-related decrease in glutamate (AMPA) receptors but no change in nerve terminal markers. *Brain Res* 589:320-326.

Bahr BA, Godshall AC, Murray, BA & Lynch G (1993) Age-related changes in neural cell adhesion molecule (NCAM) isoforms in the mouse telencephalon. *Brain Res* 628:286-292.

Bahr BA, Abai B, Gall C, Vanderklish PW, Hoffman KB & Lynch G (1994a) Induction of β-amyloid containing polypeptides in hippocampus: evidence for a concomitant loss of synaptic proteins and interactions with an excitotoxin. *Exp Neurol* 129:81-94.

Bahr BA, Lam N & Lynch G (1994b) Changes in the concentrations of tau and other structural proteins in the brains of aged mice. *Neurosci Lett* 175:49-52.

Bahr BA, Park GY, Singh M & Lynch G (1994c). Induction of aging and excitotoxicity processes in long-term hippocampal slice cultures. *Mol Biol Cell Supp* 5:468a.

Bahr BA, Tiriveedhi S, Park GY & Lynch G (1995) Induction of calpain-mediated spectrin fragments by pathogenic treatments in long-term hippocampal slices. *J Pharmacol Exp Ther* 273:902-908.

Bahr BA, Bendiske J, Brown QB, Munirathinam S, Caba E, Rudin M, Urwyler S, Sauter A & Rogers G (2002) Survival signaling and selective neuroprotection through glutamatergic transmission. *Exp Neurol* 174: 37-47.

Banik NL, Matzelle D, Gantt-Wilford G & Hogan EL (1997) Role of calpain and its inhibitors in tissue degeneration and neuroprotection in spinal cord injury. *Ann NY Acad Sci* 825:120-127.

Bartus RT, Baker KL, Heiser AD, Sawyer SD, Dean RL, Elliot PJ & Straub JA (1994a) Postischemic administration of AK275, a calpain inhibitor, provides substantial protection against focal ischemic brain damage. *J Cerebral Blood Flow Metabolism* 14:537-544.

Bartus RT, Hayward NJ, Elliot PJ, Sawyer SD, Baker KL, Dean RL, Akiyama A, Straub JA, Harbeson SL, Li Z & Powers J (1994b) Calpain inhibitor AK295 protects neurons from focal brain ischemia. *Stroke* 25:2265-2270.

Bartus RT, Dean RL, Cavanaugh K, Eveleth D, Carriero DL & Lynch G (1995a) Time-related neuronal changes following middle cerebral artery occlusion: Implications for therapeutic intervention and the role of calpain. *J Cerebral Blood Flow Metabolism* 15:969-979.

Bartus RT, Elliot PJ, Hayward NJ, Dean RL, Harbeson S, Straub JA, Li Z & Powers JC (1995b) Calpain as a novel target for treating acute neurodegenerative disorders. *Neurol Res* 17:249-258.

Bartus RT, Dean RL, Mennerick S, Eveleth D & Lynch G (1998) Temporal ordering of pathogenic events following transient global ischemia. *Brain* Res 790:1-13.

Baudry M, Bundman M, Smith E & Lynch G (1981) Micromolar calcium stimulates proteolysis and glutamate binding in rat brain synaptic membranes. *Science* 212:937-938.

Beckmann JS, Richard I, Broux O, Fougerousse F, Allamand V, Chiannilkulchai N, Lim LE, Duclos F, Bourg N, Brenguier L, Pasturaud P, Quetier F, Roudaut C, Sunada Y, Meyer J, Dincer P, Lefranc G, Merlini L, Topaloglu H, Tome FM, Cohen D, Jackson CE, Campbell KP & Fardeau M (1996) Identification of muscle-specific calpain and beta-sarcoglycan genes in progressive autosomal recessive muscular dystrophies. *Neuromuscul Disord* 6:455-462.

Bednarski E, Vanderklish P, Gall C, Saido TC, Bahr BA & Lynch G (1995) Translational suppression of calpain I reduces NMDA-induced spectrin proteolysis and pathophysiology cultured hippocampal slices. *Brain Res* 694:147-157.

Bendiske J, Caba E, Brown QB, & Bahr BA (2002) Intracellular deposition, microtubule destabilization, and transport failure: An "early" pathogenic cascade leading to synaptic decline. *J Neuropathol Exp Neurol* 61:640-650.

Blake JF, Brown MW & Collingridge GL (1988) A quantitative study of the actions of excitatory amino acids and antagonists in rat hippocampal slices. *Br J Pharmacol* 95:291-299.

Boivin P, Galand C & Dhermy D (1990) In vitro digestion of spectrin, protein 4.1 and ankyrin by erythrocyte calcium dependent neutral protease (calpain I). *Int J Biochem* 22:1479-1489.

Bronk SF & Gores GJ (1993) pH-dependent nonlysosomal proteolysis contributes to lethal anoxic injury in rat hapatocytes. *Am J Physiol* 264:G744-G751.

Butler D & Bahr BA (2006) Oxidative stress and lysosomes: CNS-related consequences and implications for lysosomal enhancement strategies and induction of autophagy. *Antioxid Redox Signal* 8:185-196.

Butler D, Brown QB, Chin DJ, Batey L, Karim S, Mutneja MS, Karanian DA & Bahr BA (2005) Cellular responses to protein accumulation involve autophagy and lysosomal enzyme activation. *Rejuvenation Res* 8:227-237.

Caba E, Brown QB, Kawasaki B & Bahr BA (2002) Peptidyl α-keto amide inhibitors of calpain block excitotoxic damage without affecting signal transduction events. *J Neuroscience Res* 67:787-794.

Capaldi D, Rosario R, Esteban ET & Bahr BA (1997) A 27-kDa matrix receptor from rat brain synaptosomes: Selective recognition of the Arg-Gly-Asp-Ser domain and unique resistance to calcium-dependent proteolysis. *Neuroscience Res* 28:275-279.

Corasaniti MT, Navarra M, Catani MV, Melino G, Nistico G & Finazzi-Agro A (1996) NMDA and HIV-1 coat protein, GP120, produce necrotic but not apoptotic cell death in human CHP100 neuroblastoma cultures via a mechanism involving calpain. *Biochem Biophys Res Commun* 229: 299-304.

del Cerro S, Larson J, Oliver MW & Lynch G (1990) Development of hippocampal long-term potentiation is reduced by recently introduced calpain inhibitors. *Brain Res* 523:313-315.

del Cerro S, Arai A, Kessler M, Bahr BA, Vanderklish P, Rivera S & Lynch G (1994) Stimulation of NMDA receptors activates calpain in cultured hippocampal slices. *Neurosci Lett* 167:149-152.

Denny JB, Polan-Curtain J, Ghuman A, Wayner MJ & Armstrong DL (1990) Calpain inhibitors block long-term potentiation. *Brain Res* 534:317-320.

Di Lisa F, De Tullio R, Salamino F, Barbato R, Melloni E, Siliprandi N, Schiaffino S & Pontremoli S (1995) Specific degradation of troponin T and I by Mu-calpain and its modulation by substrate phosphorylation. *Biochem J* 308:57-61.

Edelstein CL, Wieder ED, Yaqoob MM, Gengaro PE, Burke TJ, Nemenoff RA & Schrier RW (1995) The role of cysteine proteases in hypoxia-induced rat renal proximal tubular injury. *Proc Natl Acad Sci USA* 92:7662-7666.

Edelstein CL, Yaqoob MM, Alkhunaizi A, Gengaro PE, Nemenoff RA, Wang KKW & Schrier RW (1996) Modulation of hypoxia-induced calpain activity in rat renal proximal tubules. *Kidney Int* 50:1150-1157.

Edelstein CL, Ling H, Gengaro PE, Nemenoff RA, Bahr BA & Schrier RW (1997a) Effect of glycine on prelethal and postlethal increases in calpain activity in rat renal proximal tubules. *Kidney Int* 52:1271-1278.

Edelstein CL, Ling H & Schrier RW (1997b) The nature of renal cell injury. *Kidney Int* 51:1341-1351.

Elliget KA, Phelps PC & Trump BF (1994) Cytosolic Ca^{2+} elevation and calpain inhibitors in $HgCl_2$ injury to rat kidney proximal tubule epithelial cells. *Pathobiology* 62:298-310.

Elliott EM, Mattson MP, Vanderklish P, Lynch G, Change I & Sapolsky RM (1993) Corticosterone exacerbates kainate-induced alterations in hippocampal tau immunoreactivity and spectrin proteolysis in vivo. *J Neurochem* 61:57-67.

Fernandez-Shaw C, Marina A, Cazorla P, Valdivieso F & Vazquez J (1997) Anti-brain spectrin immunoreactivity in Alzheimer's disease: degradation of spectrin in an animal model of cholinergic degeneration. *J Neuroimmunol* 77:91-98.

Fox JEB, Taylor RG, Taffarel M, Boyles JK & Goll DE (1993) Evidence that activation of platelet calpain is induced as a consequence of binding of adhesive ligand to the integrin glycoprotein IIbIIIa. *J Cell Biol* 120:1501-1507.

Garcia-Dorado D, Rodriguez-Sinovas A, Ruiz-Meana M, Inserte J, Agulló L & Cabestrero A (2006) The end-effectors of preconditioning protection against myocardial cell death secondary to ischemia–reperfusion. *Cardiovasc Res* 70:274-285.

Govindarajan B, Laird J, Sherman R, Salomon RG & Bhattacharya SK (2008) Neuroprotection in glaucoma using calpain-1 inhibitors: regional differences in calpain-1 activity in the trabecular meshwork, optic nerve and implications for therapeutics._*CNS Neurol Disord Drug Targets* 7:295-304.

Grynspan F, Griffin WR, Cataldo A, Katayama S & Nixon RA (1997) Active site-directed antibodies identify calpain II as an early-appearing and pervasive component of neurofibrillary pathology in Alzheimer's disease. *Brain Res* 763:145-158.

Guthrie PB, Segal M & Kater SB (1991) Independent regulation of calcium revealed by imaging dendritic spines. *Nature* 354:76-80.

Harris AS, Croall DE & Morrow JS (1989) The calmodulin-binding site in α-fodrin is near the calcium-dependant protease-I cleavage site. *J Biol Chem* 263:15754-15761.

Hatakeyama T, Matsumoto M, Brengman JM & Yanagihara T (1988) Immunohistochemical investigation of ischemic and postischemic damage after bilateral carotid occlusion in gerbils. *Stroke* 19:1526-1534.

Haug LS, Ostvold AC, Cowburn RF, Garlind A, Winblad B, Bogdanovich N & Walaas SI (1996) Decreased inositol (1,4,5)-trisphosphate receptor levels in Alzheimer's disease cerebral cortex: selectivity of changes and possible correlation to pathological severity. *Neurodegeneration* 5:169-176.

Hoffman KB, Murray BA, Lynch G, Munirathinam S & Bahr BA (2001) Delayed and isoform-specific effect of NMDA exposure on neural cell adhesion molecules in hippocampus. *Neuroscience Res* 39:167-173.

Hong SC, Lanzino G, Goto Y, Kang SK, Schottler F, Kassell NF & Lee KS (1994a) Calcium-activated proteolysis in rat neocortex induced by transient focal ischemia. *Brain Res* 661:43-50.

Hong SC, Goto Y, Lanzino G, Soleau S, Kassell NF & Lee KS (1994b) Neuroprotection with a calpain inhibitor in a model of focal cerebral ischemia. *Stroke* 25:663-669.

Inomata M, Imahori K & Kawashima S (1986) Autolytic activation of calcium-activated neutral protease. *Biochem Biophys Res Communb* 138:638-643.

Inomata M, Kasai Y, Nakamura M & Kawashima S (1988) Activation mechanism of calcium-activated neutral protease. Evidence for the existence of intramolecular and intermolecular autolyses. *J Biol Chem* 263:19783-19787.

Inomata M, Hayashi M, Ohno-Iwashita Y, Tsubuki S, Saido TC & Kawashima S (1996) Involvement of calpain in integrin-mediated signal transduction. *Arch Biochem Biophys* 328:129-134.

Inserte J, Garcia-Dorado D, Ruiz-Meana M, Agullo L, Pina P & Soler-Soler J (2004) Ischemic preconditioning attenuates calpain-mediated degradation of structural proteins through a protein kinase A-dependent mechanism. *Cardiovasc Res* 64:105–114.

Jiang TX, Reid WD, Belcastro A & Road JD (1998) Load dependence of secondary diaphragm inflammation and injury after acute inspiratory loading. *Am J Respir Crit Care* 157:230-236.

Kampfl A, Posmantur R, Nixon R, Grynspan F, Zhao X, Liu SJ, Newcomb JK, Clifton GL & Hayes RL (1996) mu-calpain activation and calpain-mediated cytoskeletal proteolysis following traumatic brain injury. *J Neurochem* 67:1575-1583.

Karanian DA & Bahr BA (2006) Cannabinoid drugs and enhancement of endocannabinoid responses: Strategies for a wide array of disease states. *Current Mol Med* 6:677-684.

Karanian DA, Brown QB, Makriyannis A, Kosten TA, & Bahr BA (2005) Dual modulation of endocannabinoid transport and fatty acid amide hydrolase protects against excitotoxicity. *J Neuroscience* 25:7813-7820.

Karanian DA, Baude A, Brown QB, Parsons C & Bahr BA (2006) 3-Nitropropionic acid toxicity in hippocampus: Protection through N-methyl-D-aspartate receptor antagonism. *Hippocampus* 16:834-842.

Karanian DA, Karim SL, Wood JT, Williams JS, Lin S, Makriyannis A & Bahr BA (2007) Endocannabinoid enhancement protects against kainic acid-induced seizures and associated brain damage. *J Pharmacol Exp Ther* 322:1059-1066.

Kawasaki BT, Hoffman KB, Yamamoto RS & Bahr BA (1997) Variants of the receptor/channel clustering molecule gephyrin in brain: Distinct distribution patterns, developmental profiles, and proteolytic cleavage by calpain. *J Neuroscience Res* 49:381-388.

Kelton JG, Moore JC, Warkentin TE & Hayward CP (1996) Isolation and characterization of cysteine proteinase in thrombotic thrombocytopenic purpura. *Br J Haematol* 93:421-426.

Kluck RM, Bossy-Wetzel E, Green DR & Newmeyer DD (1997) The release of cytochrome c from mitochondria: A primary site for Bcl-2 regulation of apoptosis. *Science* 275:1132-1136.

Kochhar A, Zivin JA & Mazzarella V (1991) Pharmacologic studies of the neuroprotective actions of a glutamate antagonist in ischemia._*J Neurotrauma* 8:175-86.

Kohli V, Gao W, Camargo Jr CA & Clavien PA (1997) Calpain is a mediator of preservation-reperfusion injury in rat liver transplantation. *Proc Natl Acad Sci USA* 94:9354-9359.

Kumamoto T, Ueyama H, Watanabe S, Yoshioka K, Miike T, Goll DE, Ando M & Tsuda T (1995) Immunohistochemical study of calpain and its endogenous inhibitor in the skeletal muscle of muscular dystrophy. *Acta Neuropathol* (Berl) 89:399-403.

Kumamoto T, Ueyama H, Sugihara R, Kominami E, Goll DE & Tsuda T (1997) Calpain and cathepsins in the skeletal muscle of inflammatory myopathies. *Eur Neurol* 37:176-181.

Lee KS. The role of calcium-activated proteolysis in ischemic brain injury. In: Peterson PL & Phillis JW (eds.), "Novel Therapeutics for CNS Injuries: Rationale and Results." *Boca Raton: CRC Press,* Inc.; 1995; pp 347-361.

Lee KS, Frank S, Vanderklish P, Arai A & Lynch G (1991) Inhibition of proteolysis protects hippocampal neurons from ischemia. *Proc Natl Acad Sci USA* 88:7233-7237.

Lee KS, Foley PL, Vanderklish P, Lynch G, Goto Y & Kassell NF. The role of calcium-activated proteolysis in vasospasm after subarachnoid hemorrhage. In: Findlay JM (ed.), *"Plasticity and Pathology in the Damaged Brain: Cerebral Vasospasm."* Amsterdam: Elsevier Science Publishers BV; 1993; pp 85-88.

Lee KS, Yanamoto H, Fergus A, Hong SC, Kang SD, Cappelletto B, Toyoda T, Kassell NF, Bavbek M & Kwan AL (1997) Calcium-activated proteolysis as a therapeutic target in cerebrovascular disease. *Ann NY Acad Sci* 825:95-103.

Li Z, Patil GS, Golubski ZE, Hori H, Tehrani K, Foreman JE, Eveleth DD, Bartus RT & Powers JC (1993) Peptide α-keto ester, α-keto amide, α-keto acid inhibitors of calpains and other cysteine proteases. *J Med Chem* 36:3472-3480.

Li Z, Hogan EL & Banik NL (1996a) Role of calpain in spinal cord injury: increased calpain immunoreactivity in rat spinal cord after impact trauma. *Neurochem Res* 21:441-448.

Li Z, Ortega-Vilain A-C, Patil GS, Chu D-L, Foreman JE, Eveleth DD & Powers JC (1996b) Novel Peptidyl α-keto amide inhibitors of calpains and other cysteine proteases. *J Med Chem* 39:4089-4098.

Liu J, Liu MC & Wang KK(2008) Calpain in the CNS: from synaptic function to neurotoxicity. *Sci Signal* 1: re1

Madden KP, Clark WM & Zivin JA (1993) Delayed therapy of experimental ischemia with competitive N-methyl-D-aspartate antagonists in rabbits. *Stroke* 24:1068-1071.

Markgraf CG, Velayo NL, Johnson MP, McCarty DR, Medhi S, Koehl JR, Chmielewski PA & Linnik MD (1998) Six-hour window of opportunity for calpain inhibition in focal cerebral ischemia in rats. *Stroke* 29:152-158.

Martin SJ, O'Brien GA, Nishioka WK, McGahon AJ, Mahboubi A, Saido TC & Green DR (1995) Proteolysis of fodrin (non-erythroid spectrin) during apoptosis. *J Biol Chem* 270:6425-6428.

Masliah E, Iimoto DS, Saitoh T, Hanson LA & Terry RD (1990) Increased immunoreactivity of brain spectrin in Alzheimer disease: A marker for synapse loss. *Brain Res* 531:55-65.

Masliah E, Mallory M, Hansen L, DeTeresa R & Terry RD (1993) Quantitative synaptic alterations in the human neocortex during normal aging. *Neurology* 43:192-197.

McGinn MJ, Kelley BJ, Akinyi L, Oli MW, Liu MC, Hayes RL, Wang KK & Povlishock JT (2009) Biochemical, structural, and biomarker evidence for calpain-mediated cytoskeletal change after diffuse brain injury uncomplicated by contusion. *J Neuropathol Exp Neurol* 68:241-249.

Melloni E, Michetti M, Salamino F & Pontremoli S (1998) Molecular and functional properties of a calpain activator protein specific for mu-isoforms. *J Biol Chem* 273:12827-12831.

Mercken M, Grynspan F & Nixon RA (1995) Differential sensitivity to proteolysis by brain calpain of adult human tau, fetal human tau, and PHF-tau. *FEBS Lett* 368:10-14.

Miller ACK, Dwyer-Nield LD & Malkinson AM (1995) Very early changes in pulmonary protein kinase C-alpha and calpain II contents following injection of butylated hydroxytoluene (BHT) into mice. *Toxicology* 97:141-149.

Mondello S, Robicsek SA, Gabrielli A, Brophy GM, Papa L, Tepas J, Robertson C, Buki A, Scharf D, Jixiang M, Akinyi L, Muller U, Wang KK & Hayes RL (2010) αII-spectrin breakdown products (SBDPs): diagnosis and outcome in severe traumatic brain injury patients. *J Neurotrauma* 27:1203-1213.

Moon RT & McMahon AP (1990) Generation of diversity in nonerythroid spectrin. Multiple polypeptides are predicted by sequence analysis of cDNAs encompassing the coding region of human nonerythroid alpha-spectrin. *J Biol Chem* 265:4427-4433.

Munirathinam S & Bahr BA (2004) Repeated contact with subtoxic soman leads to synaptic vulnerability in hippocampus. *J Neuroscience Res* 77:739-746.

Nadler JV, Perry BW & Cotman CW (1978) Intraventricular kainic acid preferentially destroys hippocampal pyramidal cells. *Nature* 271:676-677.

Naidoo V, Nikas SP, Karanian DA, Hwang J, Zhao J, Wood JT, Alapafuja SO, Vadivel SK, Butler D, Makriyannis A, & Bahr BA (2011) A new generation fatty acid amide hydrolase inhibitor protects against kainate-induced excitotoxicity. *J Mol Neuroscience* 43:493-502.

Najm I, El-Skaf G, Massicotte G, Vanderklish P, Lynch G & Baudry M (1992) Changes in polyamine levels and spectrin degradation following kainate-induced seizure activity: Effect of difluoromethylornithine. *Exp Neurol* 116:345-354.

Nilsson E, Alafuzoff I, Blennow K, Blomgren K, Hall CM, Janson I, Karlsson I, Wallin A, Gottfries CG & Karlsson JO (1990) Calpain and calpastatin in normal and Alzheimer-degenerated human brain tissue. *Neurobiol Aging* 11:425-431.

Perlmutter LS, Gall C, Baudry M & Lynch G (1990) Distribution of calcium-activated protease calpain in rat brain. *J Comp Neurol* 296:269-276.

Peterson C, Vanderklish P, Seubert P, Cotman C & Lynch G (1991) Increased spectrin proteolysis in fibroblasts from aged and Alzheimer donors. *Neurosci Lett* 121:239-243.

Posmantur R, Kampfl A, Siman R, Liu J, Zhao X, Clifton GL & Hayes RL (1997) A calpain inhibitor attenuates cortical cytoskeletal protein loss after experimental traumatic brain injury in the rat. *Neuroscience* 77:875-888.

Ray SK, Samantaray S, Smith JA, Matzelle DD, Das A & Banik NL (2011) Inhibition of cysteine proteases in acute and chronic spinal cord injury. *Neurotherapeutics* 8:180-186.

Roberts-Lewis JM & Siman R (1993) Spectrin proteolysis in the hippocampus: A biochemical marker for neuronal injury and neuroprotection. *Ann NY Acad Sci* 679:78-86.

Roberts-Lewis JM, Savage MJ, Marcy VR, Pinsker LR & Siman R (1994) Immunolocalization of calpain I-mediated spectrin degradation to vulnerable neurons in the ischemic gerbil brain. *J Neurosci* 14:3934-3944.

Ryzhikov S & Bahr BA (2008) Gephyrin alterations due to protein accumulation stress are reduced by the lysosomal modulator Z-Phe-Ala-diazomethylketone. *J Mol Neuroscience* 34:131-139.

Saatman KE, Bozyczko-Coyne D, Marcy V, Siman R & McIntosh T (1996a) Prolonged calpain-mediated spectrin breakdown occurs regionally following experimental brain injury in the rat. *J Neuropathol Exp Neurol* 55:850-860.

Saatman KE, Murai H, Bartus RT, Smith DH, Hayward NJ, Perri BR & McIntosh TK (1996b) Calpain inhibitor AK295 attenuates motor and cognitive deficits following experimental brain injury in the rat. *Proc Natl Acad Sci USA* 93:3428-3433.

Saatman KE, Zhang C, Bartus RT & McIntosh TK (2000) Behavioral efficacy of posttraumatic calpain inhibition is not accompanied by reduced spectrin proteolysis, cortical lesion, or apoptosis. *J Cereb Blood Flow Metab* 20:66-73.

Saatman KE, Creed J & Raghupathi R (2010) Calpain as a therapeutic target in traumatic brain injury. *Neurotherapeutics* 7:31-42.

Saido, TC, Nagao S, Shiramine M, Tsukaguchi M, Sorimachi H, Murofushi H, Tsuchiya T, Ito H & Suzuki K (1992) Autolytic transition of μ-calpain upon activation as resolved by antibodies distinguishing between pre- and post-autolysis forms. *J Biochem* 111:81-86.

Saido, TC, Yokota M, Nagao S, Yamaura I, Tani E, Tsuchiya T, Suzuki K & Kawashima S (1993) Spatial resolution of fodrin proteolysis in postischemic brain. *J Biol Chem* 268:25239-25243.

Sato M, Tani E, Matsumoto T, Fujikawa H & Imajoh-Ohmi S (1997) Generation of the catalytic fragment of protein kinase C alpha in spastic canine basilar artery. *J Neurosurg* 87:752-756.

Seubert P, Baudry M, Dudek S & Lynch G (1987) Calmodulin stimulates the degradation of brain spectrin by calpain. *Synapse* 1:20-24.

Seubert P, Ivy G, Larson J, Lee J, Shahi K, Baudry M & Lynch G (1988a) Lesions of entorhinal cortex produce a calpain-mediated degradation of brain spectrin in dentate gyrus. I. Biochemical studies. *Brain Res* 459:226-232.

Seubert P, Larson J, Oliver M, Jung MW, Baudry M & Lynch G (1988b) Stimulation of NMDA receptors induces proteolysis of spectrin in hippocampus. *Brain Res* 460:189-194.

Seubert P, Lee K & Lynch G (1989a) Ischemia triggers NMDA receptor-linked cytoskeletal proteolysis in hippocampus. *Brain Res* 492:366-370.

Seubert P, Nakagawa Y, Ivy G, Vanderklish P, Baudry M & Lynch G (1989b) Intrahippocampal colchicine injection results in spectrin proteolysis. *Neuroscience* 31:195-202.

Seubert P, Peterson C, Vanderklish P, Cotman C & Lynch G (1990) Increased spectrin proteolysis in the brindled mouse mutation. *Neurosci Lett* 108:303-308.

Siesjö BK (1988) Historical overview. Calcium, ischemia, and death of brain cells. *Ann N Y Acad Sci* 522:638-661.

Siman R, Baudry M & Lynch G (1984) Brain fodrin: Substrate for calpain I, an endogenous calcium-activated protease. *Proc Natl Acad Sci USA* 81:3572-3576.

Siman R & Noszak JC (1988) Excitatory amino acids activate calpain I and induce structural protein breakdown in vivo. *Neuron* 1:279-287.

Siman R, Noszek JC & Kegerise C (1989) Calpain I activation is specifically related to excitatory amino acid induction of hippocampal damage. *J Neurosci* 9:1579-1590.

Small DL & Buchan AM (1997) NMDA antagonists: their role in neuroprotection. *Int Rev Neurobiol* 40:137-171.

Spencer MJ, Croall DE & Tidball JG (1995) Calpains are activated in necrotic fibers from mdx dystrophic mice. *J Biol Chem* 270:10909-10914.

Steward O, Goldschmidt RB & Sutula T (1984) Neurotoxicity of colchicine and other tubulin-binding agents: a selective vulnerability of certain neurons to the disruption of microtubules. *Life Sci* 35:43-51.

Stoppini L, Buchs P-A & Muller D (1991) A simple method for organotypic cultures of nervous tissue. *J Neurosci Meth* 37:173-182.

Suzuki K, Tsuji S, Ishiura S, Kimura Y, Kubota S & Imahori K (1981) Autolysis of calcium-activated neutral protease of chicken skeletal muscle. *J Biochem* 90:1787-1793.

Takeyama Y, Nakanishi H, Uratsuji Y, Kishimoto A & Nishizuka Y (1986) A calcium-protease activator associated with brain microsomal-insoluble elements. *FEBS Lett* 194:110-114.

Tijsen MJ, Peters SM, Bindels RJ, Van-Os CH & Wetzels JF (1997) Glycine protection against hypoxic injury in isolated rat proximal tubules: the role of proteases. *Nephrol Dial Transplant* 12:2549-2556.

Tolnadi S & Korecky B (1986) Calcium dependent proteolysis and its inhibition in ischemic rat myocardium. *Can J Cardiol* 2:442-447.

Vanderklish PW & Bahr BA (2000) The pathogenic activation of calpain: A marker and mediator of cellular toxicity and disease states. *Internatl J Exp Pathol* 81:323-339.

Vanderklish P, Saido TC, Gall C & Lynch G (1995) Proteolysis of spectrin by calpain accompanies theta-burst stimulation in cultured hippocampal slices. *Mol Brain Res* 32:25-35.

Vanderklish PW, Krushel LA, Holst BH, Gally JA, Crossin KL & Edelman GM (2000) Marking synaptic activity in dendritic spines with a calpain substrate exhibiting fluorescence resonance energy transfer. *Proc Natl Acad Sci USA* 97:2253-2258.

Vicente JS, Munirathinam S & Bahr BA (2002) Calpain-mediated spectrin breakdown identifies vulnerable brain regions at middle age. *J Mol Biol Biotech* 3:1-9.

Vornov JJ, Tasker RC & Coyle JT (1991) Direct observation of the agonist-specific regional vulnerability to glutamate, NMDA, and kainate neurotoxicity in organotypic hippocampal cultures. *Exp Neurol* 114:11-22.

Vornov JJ, Tasker RC & Park J (1995) Neurotoxicity of acute glutamate transport blockage depends on coactivation of both NMDA and AMPA/kainate receptors in organotypic hippocampal cultures. *Exp Neurol* 133:7-17.

Wang KKW & Yuen P-W (Eds.) Calpain: The Pharmacology and Toxicology of Calcium-Dependent Protease. 1st Edition. Philadelphia: *CRC Press,* Taylor & Francis Group; 448 pages (1999).

Wang L, Bhatnagar A, Ansari NH, Dhir P, Srivastava SK (1996) Mechanism of calcium-induced disintegrative globulization of rat lens fiber cells. *Invest Ophthalmol Vis Sci* 37:915-922.

Yamaura I, Tani E, Saido TC, Suzuki K, Minami N & Maeda Y (1993) Calpain-calpastatin system of canine basilar artery in vasospasm. *J Neurosurg* 79:537-543.

Yang LS, Gordon-Krajcer W & Ksiezak-Reding H (1997) Tau released from paired helical filaments with formic acid or guanidine is susceptible to calpain-mediated proteolysis. *J Neurochem* 69:1548-1558.

Yao H, Ginsberg MD, Eveleth DD, LaManna JC, Watson BD, Alonso OF, Loor JY, Foreman JH & Busto R (1995) Local cerebral glucose utilization and cytoskeletal proteolysis as indices of evolving focal ischemic injury in core and penumbra. *J Cereb Blood Flow Metab* 15:398-408.

Zhao X, Pike BR, Newcomb JK, Wang KK, Posmantur RM & Hayes RL (1999) Maitotoxin induces calpain but not caspase-3 activation and necrotic cell death in primary septo-hippocampal cultures. *Neurochem Res* 24:371-382.

In: Enzymes and Enzyme Activity
Editor: Eva M. Lashinski
ISBN: 978-1-62417-696-8

Chapter III

Calpain: Structure, Biology and Clinical Significance

Muthuraman Pandurangan and Inho Hwang
Dept. of Animal Science,
Chonbuk National University, Jeonju, South Korea

Abstract

Calpains are class of proteins which belongs to the calcium dependent, non-lysosomal cysteine proteases. There are three major types of calpains expressed in muscle namely, μ-calpain, m-calpain and calpain 3, which shows proteolytic activities. Skeletal muscle fibres that possess all the three calpains also contain Ca^{2+}-dependent proteases. The functional role of calpains was found to be associated with apoptosis and myogenesis. However, calpain 3 is likely to be involved in sarcomeric remodeling. A defect in the expression of calpain 3 results in limb-girdle muscular dystrophy type 2A. Calpain 3 is bound in skeletal muscle fibres at the N2A line of the large elastic protein, titin. A substantial proportion of calpain 3, but not ubiquitous calpains, is activated 24h after a single bout of eccentric exercise. In vitro studies indicated that calpain 3 can be activated to 2-4 folds higher than normal resting cytoplasmic [Ca^{2+}] if exposed for a prolonged period of time and this suggests sustained increase in intracellular [Ca^{2+}]. Characterization of calpain system in developing muscle is needed to indicate which calpastatin isoforms are present and whether both μ- and m-calpain exist in differentiating myoblasts. It has been demonstrated that the activation of ubiquitous calpains and calpain 3 in skeletal muscle is very well regulated in the presence of huge and rapid changes in intracellular [Ca^{2+}].

Diabetes mellitus is recognized as a clinical syndrome that is characterized by hyperglycemia due to deficiency of insulin. Non-insulin dependent diabetes mellitus (NIDDM/Type-II), is multifactorial and exact biochemical and genetic defects have not been elucidated completely. The calpains seems to play a role in NIDDM and positional cloning experiments revealed NIDDM susceptibility to calpain 10 (CAPN 10). The increased calpain activity and leukocyte trafficking were noticed in the microcirculation of ZDF (Zuker diabetic fatty) rats. The exercise and low body weight play significant roles on the reducing the calpain expressions or elevating the calpains degradation in the skeletal muscle of NIDDM rats. Numerous investigations reported that non-coding polymorphisms in CAPN10 proteins may involved in the NIDDM. The calpain and its mRNA presence have been reported in tissues from many mammalian species while CAPN10 and other calpains seems to be linked with glucose metabolism, insulin secretion and action pathways.

Keywords: calpains; calpastatin; skeletal; protein; proteases

Introduction

Calpains are class of proteins which belongs to the Ca^{2+} dependent, non-lysosomal cysteine proteases [1]. There are three major types of calpain expressed in skeletal muscle namely, μ-calpain, m-calpain and calpain 3 (p94) and shows proteolytic activities. μ-calpain and m-calpain are calcium activated proteases and requires micro and millimolar concentration of Ca^{2+} for activation, respectively [2]. They are ubiquitously expressed in the skeletal muscle. Calpain 3 requires very little or no Ca^{2+} for its activity and are characterized by rapid and complete autolysis [3]. In addition to calpains, calpain inhibitor, calpastatin is also found in the skeletal muscle which is a highly polymorphic protein and calpain-specific endogenous inhibitor. They are found to be associated with calpain proteolytic enzyme family [4]. It was originally proposed that calpains are responsible for initiating metabolic turnover of myofibrillar proteins and therefore affects the rate of muscle protein degradation [5]. Recent investigations have shown that calpain activity is mandatory for myoblast fusion and for cell proliferation in addition to cell growth [6]. Hence, calpains also affect the number of skeletal muscle cells in domestic animals by changing rate of myoblast proliferation and myoblast fusion modulation. In this review we will focus on the role of calpains system in skeletal muscle.

Calpain history was started in 1964, when they were detected in brain, eye lens. Calpains activities were caused by intracellular cysteine proteases and having an optimal activity at neutral pH [7]. Calpains activity was found to be attributable to two main isoforms, μ-calpain and m-calpain. These two isoforms are differed in their calcium requirements in vitro and their names displayed that they are activated by micro and millimolar concentrations of Ca^{2+} within the cells [8]. The μ-calpain and m-calpain terms were first used in 1989 to address the requirement of micromolar and millimolar amount of Ca^{2+} respectively for their activation and they are referred to as ubiquitous calpains [9].

Calpain 3 is a third member of the calpain family and it was discovered thirteen years ago that mutations in calpain 3 result in an autosomal recessive and progressive form of limb girdle muscular dystrophy called limb girdle muscular dystrophy type 2A [10]. Calpastatin was discovered during purification of m-calpain from the muscle extracts that shows both calpain and calpain inhibitor activity [11]. The calpastatin name was given by Takashi Murachi. The calpastatin system is involved in membrane fusion events such as neural vesicle exocytosis and platelet and red-cell aggregation. Calpastatin inhibits both μ- and m-calpain but not calpain 3 inhibited and it may regulate the activity of μ- and m-calpain by degrading calpastatin [12].

Calpain Structure

The μ-calpain and m-calpain, which are named on the basis of calcium requirement for their activity *in vitro* and required a neutral pH for their proteolytic (Figure 1). The μ-calpain and m-calpain are heterodimers containing of a catalytic and a regulatory subunit activity [13, 14]. The regulatory subunit is similar to both calpains and coded by *CAPNS1.* The catalytic subunits differ between μ-calpain and m-calpain and are coded by calpain 1 and calpain 2 respectively [15, 16]. The catalytic subunits have four domains (DI - DIV) and regulatory subunits have two domains (DV and DVI) respectively. DI is autolysed during calpain activation by calcium. DII is the conserved protease domain, which is sub-divided into the subdomains such as IIa and IIb, which form a signal domain during calcium binding. DIII consisting C2 domains, which is participated in structural changes during calcium binding [17, 18]. The carboxy-terminal domains DIV and DVI contain five EF hands and the fifth EF hand involves the subunits dimerization. DVI of the regulatory subunit consisting glycine residues that may interact with the plasma membrane and autolysed during calpain activation [19, 20].

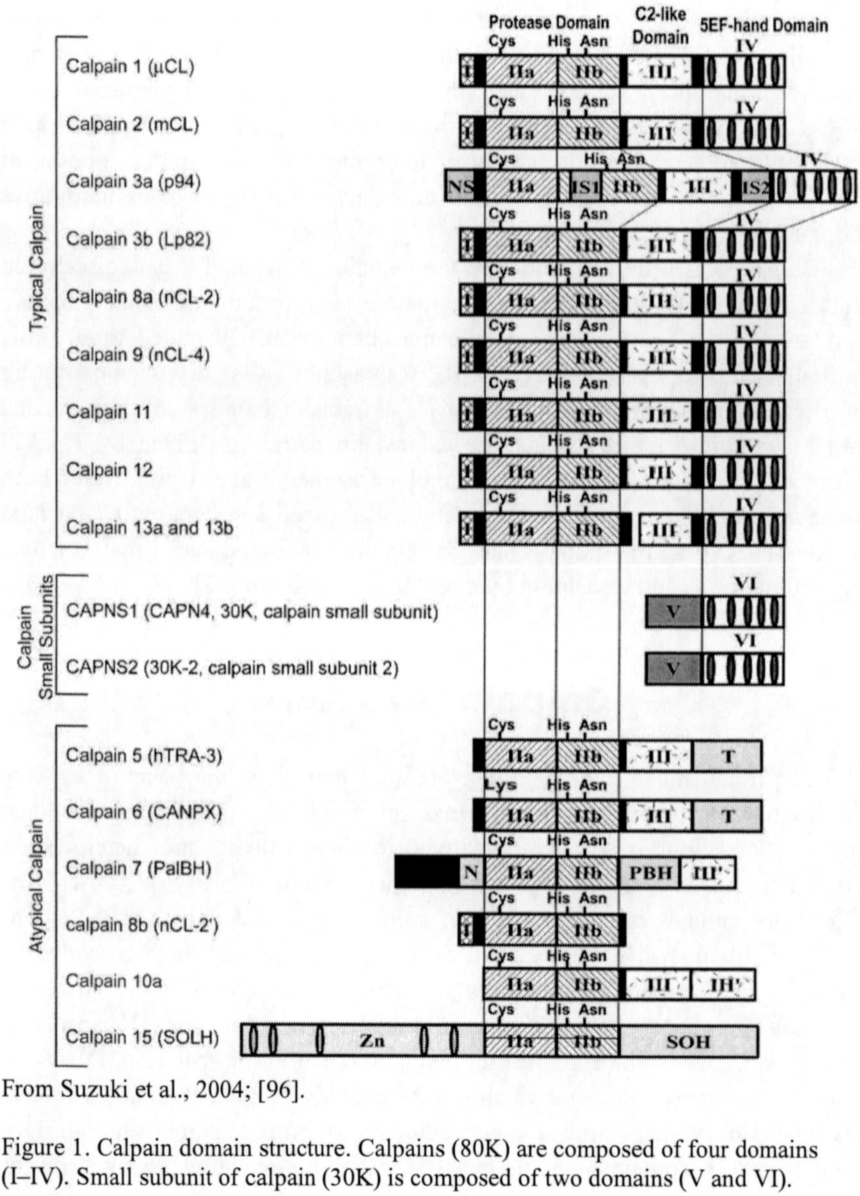

From Suzuki et al., 2004; [96].

Figure 1. Calpain domain structure. Calpains (80K) are composed of four domains (I–IV). Small subunit of calpain (30K) is composed of two domains (V and VI).

Calpain Activation

Several of mechanisms activate calpain by reducing the calcium requirements such as autolysis of the DI domain and interaction with membrane phospholipids [21]. The μ-calpain and m-calpain can be phosphorylated by protein kinase Cι and this phosphorylation is linked with elevated cell migration and invasion of lung cancer cells [22]. Extracellular signal-regulated kinase (ERK) phosphorylates m-calpain on a specific serine residue, influencing cell adhesion, motility, reduction on calcium concentration and this is required for activation of m-calpain [23]. Localization of m-calpain at the plasma membrane via PIP_2 anchorage, is required for protease activation. As well as, m-calpain may co-localize in calveolae to enhance compartmentalized signaling [24, 25]. The experiment results indicated that μ-calpain would be found in the intermembrane mitochondrial space. The calpain's influence on cellular processes, the localization and their distribution is expected to modify cellular output followed by activation of calpain [26, 27].

Calpains Regulation

Calpastatin is the ubiquitously expressed endogenous inhibitor of μ-calpain and m-calpain. The inhibitory action of calpastatin requires calcium-induced structural changes in calpain to permit the A, B and C regions within each inhibitory domain of calpastatin to bind and region B blocks the calpain active site [28]. Calpastatin is released into cytosol from intracellular storage aggregates following calcium influx to permit interaction with calpain [29].

Role of Calpains in Muscle

There is much ample evidence that the calpains have main roles, both in normal and postnatal skeletal muscle growth, muscle wasting and loss of muscle mass. A number of experiments have reported that the calpain system also has a role in normal skeletal muscle growth. Administration of b-adrenergic agonists to animal results in a 10–30% elevation in rate of accumulation of muscle mass [30]. Even though various experiments were carried out in various species with varying b-adrenergic agonists, different conditions produced marginally different results. The most studies agree that administration of b-adrenergic agonists elevates both the rate and the efficiency. It has been reported that the b-adrenergic agonists administration affects the activity of calpain system [31] and elevates muscle calpastatin activity [32]. This elevation ranges from 52 to 430% [33]. Muscle μ-calpain activity is either reduced or unchanged, while m-calpain activity seems to be elevated [34]. It has been reported that the activity of both μ- and m-calpain is reduced. The skeletal muscle mass was 30–40% greater than in half-siblings not expressing the callipyge trait in the callipyge phenotype sheeps [35]. Calpastatin activities in the affected muscles from callipyge lambs are 68–126% greater than in the same muscles from normal lambs, while calpastatin activities in the unaffected callipyge muscles and these findings suggest that elevated rate of skeletal muscle growth can result from a reduction in rate of muscle protein degradation. This reduced rate of muscle protein degradation is associated with the reduction in the activity of calpain system [36].

Recent investigation reports that the calpain activity is necessary for cells to progress through the G1 to S phase of the mitotic cycle and also required for myoblast fusion [6]. This suggests that enhanced calpain activity during muscle development may be associated with an elevated number of myoblasts. Growing C2C12 myoblasts contain m-calpain only [37]. This is due to the administration of b-adrenergic agonist to growing animals which seems to have less effect on m-calpain than on calpastatin.

Calpains are believed to play a role in protein modification with involving limited proteins degradation and consequent targeting to the ubiquitin proteasome degradation pathway [38]. The μ-calpain undergone autolysis from the full-length 80 kDa form to the 78 kDa form by increasing its Ca^{2+} sensitivity greater than 10-fold. The pre-activated μ-calpain under the surface membrane and it has been found that in the presence of micromolar Ca2+ the freely diffusible, full-length μ-calpain binds quickly at least in part to cell membranes and stays bound for some time. The Ca2+ sensitivity of μ-calpain autolysis elevated in the presence of phospholipids. The pre-activated calpain ideally placed for facilitating rapid membrane resealing, a process known to be dependent on dysferlin and local remodeling of the cytoskeleton by calpain [39]. In vivo, μ-calpain has been shown to play a role in membrane repair and it demonstrated the supports of calpains protective role in skeletal muscle.

Many reports have indicated that the calpains have very limited and specific effects on skeletal muscle proteins. Myosin is degraded very slowly and even this slow degradation is limited to a few cleavages of the light chains. Non-denatured actin is not cleaved by either μ- or m-calpain. Both μ-

and m-calpain rapidly cleave troponin T, desmin, vinculin, talin, spectrin, nebulin, and titin; more slowly cleave troponin I, filamin, C-protein, dystrophin, and tropomyosin; very slowly cleave alpha-actinin and M protein [30].

Incubation of skeletal muscle proteins with calpains results in the loss of N2 line and then in complete loss of Z-disks, leaving a gap in the middle of sarcomere. Simultaneously, loss of periodicity in the I-band area occurs due to troponin and tropomyosin degradation. The loss of Z-disk structure is caused by release of alpha-actinin from Z-disks in a nearly intact form [40]. Even though many experiments have indicated that myosin, actin, and alpha-actinin are degraded by the calpains, it is unclear whether these investigations have utilized non-denatured proteins. Both μ- and m-calpain rapidly degrades denatured actin, myosin, and alpha-actinin. The calpains specifically cleave several cytoplasmic proteins which includes mostly protein kinases and phosphatases. They do not cause bulk degradation of sarcoplasmic proteins into small fragments [41]. Most of the experiments attempting to determine muscle protein turnover, use release of free amino acids or 3-methyl histidine to estimate the rate of muscle protein degradation [40].

The myofibrillar structure must remain intact for skeletal muscle cells to be functional. The turnover of myofibrillar proteins, which represent the major fraction of total protein in muscle cells must proceed through a different mechanism. Even though the mechanism by which myofibrillar proteins turnover is still not clear and remains an area of active research, presently it seems that this turnover proceeds via two steps: 1) removal of the proteins from the myofibrillar structure and this must occur without disruption of the myofibril which extends continuously from one end of the muscle cell to the other. 2) Degradation of the individual myofibrillar proteins to small peptides and free amino acids.

Kinetic investigations have reported that proteins in adult cardiac myocytes are degraded from two different pools: 1. 10% of total muscle protein turns over rapidly with a mean half-life of 11.9 h and 2. Remaining 90% of the total muscle protein turns over more slowly with a half-life of 15.6 d. Similar experiments have not been done on skeletal muscle cells, but the structure of cardiac myofibrils, which constitute a smaller percentage of total muscle protein (45–55%) than in skeletal muscle, is very similar to the structure of skeletal muscle myofibrils and the myofibrillar proteins probably turn over via the same mechanism in two types of cells. The presence of proteins turning over at different rates in muscle cells does not prove that the myofibrillar proteins must be removed from the myofibril before they can be degraded to amino acids [40].

Many experiments reported that the 5–15% of total myofibrillar protein in skeletal and cardiac muscle cells can be dissociated from intact myofibrils in the form of myofilaments by using gentle agitation in an ATP-containing solution [42]. These easily releasable filaments lack alpha-actinin, desmin, titin, and other cytoskeletal proteins such as filamin having molecular masses above 200 kDa but contain the major myofibrilar proteins, actin and myosin. Proteins in easily releasable filaments fraction turn over rapidly, indicating that they are in a pool of rapidly degraded proteins. The easily releasable filament levels in muscle, elevates in response to treatments that increase calpain activity and decrease in response to treatments that inhibit calpain activity [43, 44].

The calpains make specific cleavages in cytoskeletal proteins that are involved in maintaining the myofibrillar structure: a) degradation of desmin, vinculin, talin, dystrophin, and spectrin. b) loss of Z-disks and release of alpha-actinin evidently due to cleavage of the N-terminal end of the large titin polypeptide and to the rapid cleavage of nebulin. [45]; c) degradation of C-protein, which encompasses thick filaments like staves around a barrel. d) degradation of troponin T and tropomyosin, which would contribute to "weakening" of the thin filament. e) degradation of M proteins, which would, along with the degradation of C protein. The calpains are unique among proteolytic enzymes in that they do not rapidly degrade the major muscle proteins, actin and myosin, which also are the major components of the easily releasable filaments. Large size and ordered structure of intact myofibrils would prevent them being taken up into lysosomes [46]. Neither myofibrils nor structurally recognizable fragments of myofibrils have been observed in lysosomal structures. The proteasome has no effect on myofibrillar proteins when they are in the myofibrillar structure [47, 48].

The available evidence suggests that the myofibrillar proteins in skeletal muscle cells are first removed from the myofibril either in the form of filaments or as individual protein. These protein molecules in the filaments are then degraded to amino acids by proteolytic systems in the cell cytoplasm. Some experimental results have suggested that individual proteins are removed from myofibrils. The mechanism by which this exchange is accomplished without disrupting the myofibrillar structure is not known. The presence of easily releasable filaments, suggests that myofibrillar proteins are released from the surface of myofibrils. This mechanism would leave functionally intact, although smaller, myofibrils as turnover progressed.

The removed myofilaments could either reassemble back onto the surface of the myofibril or be degraded to amino acids/small peptides by cytoplasmic proteinases [30].

Calpains would initiate disassembly of the myofibril by specific cleavages of Z-disk proteins at the surface of myofibril, releasing the thin filaments from their attachments to the myofibril. The myosin thick filaments attached to the released thin filaments would dissociate in the presence of ATP in the cell. Calpain-induced cleavage of C-protein and M-protein would lead to further dissociation of thick filaments to individual myosin molecules and these are degraded by proteasome or taken up into lysosomes and degraded by lysosomal cathepsins. Calpain-induced cleavage of tropomyosin and troponin T and I together with degradation of nebulin would favor dissociation of thin filaments to actin monomers and these are degraded by the proteasome.

The disassembly and degradation process is teleologically attractive and proposes definite roles for calpains and proteasome that are consistent with the known properties of these systems. An experiment done over 35 years ago suggested that newly synthesized myofibrillar proteins were added to the surface of growing myofibrils. This mechanism also indicates that the interior of myofibrils would be immortal unless the entire myofibril was turned over. Even though, there is remarkable circumstantial evidence indicating that the calpains play a main role in initiating turnover of the myofibrillar proteins, it is clear that the calpains cannot degrade the myofibrillar or any other class of proteins to amino acids. Consequently, at least two or three proteolytic systems are involved in turnover of the myofibrillar proteins. Calpain cleavage frequently ablates the regulation that normally governs activity of these kinases/phosphatases, and leaves constitutively active enzymes. The effects of active enzymes on signal transduction are not clear and it is ironic that the clearly defined property of the calpains does not involve a function in living cells, but rather its role in postmortem tenderization [49].

Calpains have already been demonstrated to participate in the turnover of several other muscle cytoskeletal proteins under different conditions.For example, desmin and dystrophin were shown to be calpain targets in cultured myotubes during periods of serum withdrawal [50] Similarly, desmin, talin, and fibronectin were shown to be cleaved by calpain during membrane cytoskeleton rearrangement concomitant with myoblast fusion [51].Recent evidence corroborating the idea that calpain regulates physiological muscle protein turnover was obtained in transgenic mice. In this study the muscle-specific over expression of a calpastatin transgene resulted in significant muscle hypertrophy and accumulation of several proteins, presumably as a consequence of reduced degradation. Preactivated calpain seems ideally placed for facilitating rapid membrane resealing if there is localized damage, a process known to be dependent on dysferlin and local remodeling of the cytoskeleton by calpain [38].

An huge amount of evidence acquired during the past 35 years has indicated that the calpains are responsible for up to 95% of all proteolytically induced postmortem tenderization that occurs during the first 7-14 d of postmortem storage at 2- 4°C. Storage for longer durations postmortem involves some catheptic proteolysis and it is still unknown whether postmortem proteolysis involves primarily μ- or m-calpain or both [49]. Experiments on animals that have received b-agonists or animals having a *Bos indicus* have indicated less tender meat. Muscle calpain activity, seems to vary little among these different groups of animals. High calpastatin activity decreases the ability of calpains to degrade myofibrillar proteins during postmortem storage. This is because of the muscle calpastatin levels vary more widely in response to different treatments than muscle calpain activities.

Even though the physiological function of μ-calpain is not clear, along with the other ubiquitously expressed calpain, m-calpain, it has been implicated in a number of cellular functions linked to Ca^{2+} signaling, including apoptosis, myogenesis, cell signaling and cell differentiation [52]. Calpains are believed to play a role in protein modification with one course of action involving limited degradation of proteins and consequently targeting to the ubiquitin proteasome degradation pathway [53]. It has been suggested that a specific role of m-calpain in skeletal muscle is to prevent major degradation by reducing Ca^{2+} release from the sarcoplasmic reticulum, following periods of excessive Ca^{2+} release [54]. This intervention seems to occur through interference with excitation-contraction matched in which the communication at the triad junction between the voltage sensors and the calcium release channels is hindered. The m-calpain plays a role in membrane repair in in vivo [55]. μ-calpain is inhibited by endogenously expressed calpastatin [38].

Following exposure to micromolar [Ca^{2+}], full-length 80 kDa μ-calpain autolysis to its 78 kDa and 76 kDa isoforms. The presence of autolysed μ-calpain is an indicator of an activated protease. Biochemical assays have demonstrated that the 80 kDa isoform of μ-calpain can itself be active if the [Ca^{2+}] is sufficiently enough [2]. Autolysed μ-calpain isoforms have a huge Ca^{2+} sensitivity than the full-length 80 kDa isoform. The skeletal muscle μ-calpain was active at a similar rate in the presence of either 20 μM Ca^{2+}or 2 μM Ca^{2+} . Above a baseline level of autolysis in human skeletal muscle, there was an additional 23%, 28% and 46% autolysis following 1 min exposure to 2.5, 10 and 25 μM Ca^{2+}, respectively [56]. Calpains participates in the physiological turnover of utrophin in muscle cells, particularly in dystrophin-deficient muscle cells where impaired Ca^{2+} homeostasis has been reported. A major advance created in this study compared to previous biochemical based studies is that the μ-calpain was expressed endogenously in skeletal muscle and was examined in the presence of all other muscle constituents, including factors such as DUK114 [57].

The role of calpain-3 in muscle repair and maintenance in myogenesis and apoptosis has been proposed.

The absence of calpain-3 in mouse skeletal muscle produces an overt phenotype [58]. Muscular dystrophy with myositis is a mouse model of the human disease human tibial muscular dystrophy and limb-girdle muscular dystrophy type 2A [59]. Calpain-3 has been shown to regulate nuclear IkBak [60]. Calpain-3 is inactive in full form and active by removing by IS1 region, process is Ca^{2+}dependent. The cleaved domains must remain associated with each other for the protease to be active [61]. Calpain-3 was found to be spontaneously autolysed in vivo. Calpain-3 is quite stable in skeletal muscle and not inhibited by calpastatin. The specific endogenous substrates for calpain-3 not identified yet. The identification the substrates for calpain-3 are important to define the role of calpain-3 in skeletal muscle. Many proteins have been identified as substrates for calpain in vitro such as filamin, vinexin, ezrin and talin and myosin light chain [62].

The intracellular [Ca^{2+}] often changed in skeletal muscle and there was a large transient elevation in intracellular [Ca^{2+}] with cytoplasmic [Ca^{2+}] reaching 2-20 mM during normal contraction (Chin and Allen, 1996). Mild and prolonged elevation in intracellular [Ca^{2+}] duration fatigue such as following eccentric exercise and there was little persistent increase in resting intracellular [Ca^{2+}] in certain disease states and aging. The cytoplasmic [Ca^{2+}] is enhanced 2-3 fold above resting levels for till 48 h following eccentric exercise in mouse skeletal muscle.

There are two mechanisms involved in calpain-3 activation following an eccentric bout of exercise. Firstly the muscle fibers are lengthened during muscle contraction. Secondly, the intracellular [Ca^{2+}] would be expected to remain elevated slightly above normal resting cytoplasmic levels for 24-48 h following the exercise [63]. Calpain-3 activated if its IS2 region was not bound to titin and its activation might be sensitive to a muscle fiber stretching [46].

Biology and Clinical Significance

Calpains in Muscle Pathology

In the muscular dystrophies and other muscle pathologies the intracellular free Ca^{2+} concentration is enhanced. This increased intracellular Ca^{2+} stimulates calpain activity. The structural alteration observed in the atrophying muscles [33] is mimicked by treatment of myofibrils from normal muscle with μ-calpain or m-calpain [36]. SDS-PAGE has reported that the calpains cause similar degradative changes in myofibrils as those seen in Duchenne muscular dystrophy.

Calpains in Human Diseases

The structural and functional diversity of calpains in the cell is displayed at their involvement in the disease pathogenesis and there are two well known genetic disorders such as cancer and limb-girdle muscular dystrophy type 2A [64, 65].

The calpain 10 is identified as a susceptibility gene for NIDDM and calpain 9 is identified as a tumor suppressor for gastric cancer. Moreover, the calpains hyperactivation is involved in several pathologies associated with changed calcium homeostasis such as Alzheimer's disease and cataract formation [66].

The calpains role also found in secondary degeneration resulting from acute cellular stress following myocardial ischemia, cerebral ischemia, traumatic brain injury and spinal cord injury. The calpains are extensively activated due to Ca^{2+} influx following cerebrovascular accident. Elevated calcium concentration in the cells results in calpain activation, which leads to uncontrolled proteolysis of both target and non-target proteins as well as consequent irreversible tissue damage.

Enormous amount of active calpain which degrades cytoskeleton molecule such as spectrin, microtubule subunits, microtubule-associated proteins, and neurofilaments and It can also damage ion channels, cell adhesion molecules, other enzymes and cell surface receptors [67, 68, 69]. This will leads to cytoskeleton and plasma membrane degradation. Calpain may also break down sodium channels leading to an influx of sodium into the cell, which results, the neuron's depolarization and more Ca^{2+} influx [70].

Calpains and Non-Insulin Dependent Diabetes Mellitus (IDDM)

The calpain-10 is the first diabetes gene to be identified through a genome scan [71]. The experimental results have reported that the variation in *CAPN10*, affects susceptibility to NIDDM in Mexican Americans, the Botnia region of Finland and the German population of Saxony. The inheritance of a specific haplotype combination defined by three single-nucleotide polymorphisms (SNP), SNP-43, -19, and -63 was found to be associated with a threefold increased risk and these polymorphisms are in non-coding region of *CAPN10. It is* believed that it alter risk by affecting the transcriptional regulation of calpain-10 [72]. SNP-43 was also found to be linked with measures of insulin action in Pima Indians with normal glucose tolerance, indicating that calpain-10 enhances susceptibility to NIDDM via its effects on the oxidation of glucose in skeletal muscle [73].

SNP-19 variant and the 111 haplotype contribute to the risk of NIDDM in Tunisian subjects and no significant link between *CAPN10* diplotypes and NIDDM were reported for Tunisians [74]. Even though, haplotype of SNPs-44, -43, -19 and -63 may be associated with NIDDM and none of these SNPs may be individually associated with NIDDM [75]. SNP-43 polymorphisms of calpain-10 have associations with increased levels of total cholesterol in NIDDM [76]. Subjects with NIDDM, the frequency of missense mutations in the insulin-promoter–factor 1 gene is greater in younger individuals than older individuals [77, 78]. SNP-44 may either directly change susceptibility to NIDDM or be in linkage disequilibrium with the disease pre-disposing variant.

Functional analysis reported that SNP-44 plays a role in the transcriptional regulation of *CAPN10*. Even though, the polymorphisms and haplotypes linked with NIDDM vary between populations and this may be due to the presence of multiple susceptibility alleles at *CAPN10* and different patterns of linkage disequilibrium between these polymorphisms and a common causal variant [72]. The amino acid polymorphisms identified in the U.K. populations and these amino acid polymorphisms were absent in the Mexican American population. The amino acid polymorphisms may change the risks linked with haplotypes and haplotype combinations in the Mexican American population. The *CAPN10* in the U.K. population did not provide replication of the polymorphisms or haplotypes at *CAPN10* that are associated with the increased risk of NIDDM in the Mexican American, Finnish or German populations. Even though, they do provide replication at the level of the locus,

CAPN10, with various alleles of susceptibility gene being associated with elevated risk in the U.K. population [72].

The calpain 3 seems to be functionally linked with CAPN10, which leads to the speculation of this enzyme that may play a role in NIDDM. Recent investigation indicated that reduced expression of CAPN3 in skeletal muscle was associated with obesity and insulin resistance and linked with NIDDM [79]. The disease develops as a conspiracy between the environment and the genetic background. The common variants in a number of genes are associated with NIDDM recently. Even though, the first gene to be linked to NIDDM by positional cloning was calpain-10 [72] and polymorphisms in this gene have been associated with NIDDM in some populations [80, 81, 82, 83]. Calpain-10 has been suggested to influence both insulin secretion and resistance As well as, common variants in the *CAPN10* gene is associated with transcript levels in skeletal muscle and adipose tissue [84, 85, 86, 87, 88].

The calcium plays an important role in insulin release in response to secretagogues and since calpain-10 has been implicated in the pathogenesis of NIDDM. The calpain-10 has been suggested to play a role in facilitating the actin reorganization required for glucose-stimulated insulin release and inhibition of calpains blocks insulin secretion [88]. Calpain-10 was identified in human pancreatic islets and there is a positive correlation between calpain-10 levels and insulin release in response to a secretagogue cocktail in INS-1 cells. Calpain-10 is involved in the first phase exocytosis and it binds to the plasma membrane SNARE complex and affect SNAP-25 proteolysis [89]. Surprisingly, calpain-10 has been suggested to induce apoptosis in response to fatty acids and therefore, calpain-10 has a positive effect on insulin secretion of healthy islets, while it may enhance lipotoxic apoptosis activation in NIDDM.

Calpain 10 enzyme expression is changed in pancreatic, muscle, and fat cells and is related with abnormal glucose homeostasis via the modification of calpain 10-dependent pathway. Calpain-10 mRNA low level in skeletal muscle were also associated with insulin resistance [73]. These results indicated that the related genetic expression of calpain-10 may play a vital role in determining the normal response of skeletal muscle to the insulin effects. The pharmacological inhibition of calpain may cause insulin resistance of glucose transport in skeletal muscle. The m-calpain may degrade glucose transporter turnover, which is the indication of calpains's role in the adjustment of muscle glucose equilibrium and muscle mass [90]. The calcium-dependent protease μ-calpain is elevated in the diabetic vasculature. The inhibition of calpain activity weakens the vascular dysfunction linked with chronic diabetes, which is the clear evidence of calpain role in the pathophysiology of diabetic vascular disease [91].

The inappropriate activation of calpain changes platelet function, partially degrades a spectrum of proteins and results in hyper-agreeability. Changes in the activity of calpain in different cells involved in diabetes-related pathways. The calpain causes endothelial dysfunction and vascular inflammation in the microcirculation of the ZDF rat, a genetic rat model of NIDDM. Some experimental results indicated that the increased calpain activity and leukocyte trafficking in the microcirculation of ZDF rats and there was inhibition of calpain activity significantly attenuated leukocyte-endothelium interactions in the vasculature of ZDF rats. The cell adhesion molecules expression increased in the vascular endothelium of ZDF rats. Moreover, the calpain inhibition may suppress the adhesion molecules between leukocyte and endothelium [91].

NIDDM-linked SNP-43*G-allele is linked with low levels of calpain 10 mRNA when compared to SNP-43*A-allele. Insulin and age had no significant effect on calpain 10 mRNA levels [92]. The exercise reduced the both μ- and m-calpain mRNA expressions, suggesting that the exercise may be related to the degradation of calpains in NIDDM rats. The changed calpain system following exercise might influence and regulate the glucose transport related pathways in skeletal muscle. Even though the changes in body weight and the μ-calpain and m-calpain mRNA expressions in skeletal muscle, exercises improve the physiopathology of diabetic subjects and some experimental results indicated that calpains as possible markers for prognoses NIDDM [93].

The expressions of calpain 10, μ-calpain, and m-calpain in obese rats were greater than in lean rats and these results indicated that both exercise and low body weight play significant roles on the reducing calpains expressions or elevating the calpains degradation and both were linked with glucose transport in the skeletal muscle of NIDDM rats. The increased body weight might increase the calpain mRNA levels, while the exercise might reduce the calpain mRNA expression. The expressions of calpains in skeletal muscle are regulated by exercise training in both lean and obese Zucker rats, which indicates that exercise can improve the glucose uptake into the contracting skeletal muscle in this model of NIDDM and the greater μ-calpain and m-calpain mRNA expressions might be risk factors for NIDDM.

The free fatty acids activate protein kinase C, which leads to hyperphosphorylation of insulin receptors, leading to reduction in the kinase activity of insulin receptors and thus enhancing insulin resistance. The down-regulation of protein kinase C activity seems to be an important factor to maintain proper phosphorylation levels of insulin receptors [94, 95]. Polymorphism within intron 3 of *CAPN10* affects mRNA expression and those with the G/G genotype have reduced calpain mRNA levels and a reduced level of calpain leads to up-regulation of protein kinase C activity, reduces insulin signaling and results in insulin resistance [73].

Acute exposure of mouse pancreatic islets to calpain inhibitors increased glucose-induced insulin secretion but decreased insulin-stimulated glucose uptake into adipocytes and skeletal muscle and glycogen synthesis is reduced in muscle. The calpain inhibition showed reduced effects on a hypoxia-associated elevation in glucose uptake and this is not mediated by the insulin receptor system [84]. The PD147631 is another calpain inhibitor, which interacts with calpain at domain IV, showed similar effects and suggests that the observed effect of calpain is not due to inhibition of calpain 10. Experiments on the action of calpain using calpain inhibitors indicates that calpain 10 and other calpains also participate in insulin secretion, action and susceptibility to NIDDM. Therefore, calpains might become new molecular target for the prevention and treatment of NIDDM and its complications [91]. The detailed experiments are needed to understand the molecular and physiological mechanism explaining the link between calpain with NIDDM.

Conclusion

Skeletal muscle fibres contain m-calpain, μ-calpain, muscle-specific calpain-3 and Ca^{2+}-dependent proteases. Their physiological roles are unclear. Even though ubiquitous calpains have been associated with apoptosis and myogenesis, calpain-3 is likely to be involved in sarcomeric remodeling. A defect in the expression of calpain-3 results in limb-girdle muscular dystrophy type 2A.

At resting intracellular [Ca^{2+}], calpains are presented in full-length in unautolysed forms. Calpain-3 is bound in skeletal muscle fibres at the N2A line of the large elastic protein, titin. Ubiquitous calpains or calpain-3 is activated immediately following sprint, endurance despite the frequent episodes of high cytoplasmic [Ca^{2+}]. A substantial proportion of calpain-3, but not ubiquitous calpains, is activated 24h after a single bout of eccentric exercise. *In vitro* studies indicated that calpain-3 becomes activated to 2-4 folds higher than normal resting cytoplasmic [Ca^{2+}] if exposed for a prolonged period of time and this suggests the sustained increase in intracellular [Ca^{2+}].

Measurement of calpastatin isoforms present in different muscles would show whether these isoforms change in response to treatments can alter muscle growth. Moreover, this would indicate whether the relationship between calpastatin activity and rate of skeletal muscle growth in mature animals is the result of a change to a calpain isoform with different calpain inhibitory properties.

Characterization of the calpain system in developing muscle is needed to indicate which calpain isoforms are present and whether both μ- and m-calpain exist in differentiating myoblasts. It has been demonstrated that the activation of ubiquitous calpains and calpain-3 in skeletal muscle is very well regulated in the presence of huge and rapid change in intracellular [Ca^{2+}].

Diabetes mellitus represents a major medical issue affecting millions of people all over the world [83]. This manuscript clearly indicates that there is a huge evidence indicating a functional association between NIDDM and genetic variations in the calpain 10 gene and other calpains. The experimental results reports that calpain 10 and other calpains are involved in metabolic pathways associated with NIDDM and calpain inhibitors able to strongly affect glucose metabolism, insulin action and insulin resistance. The increased calpain activity and leukocyte trafficking were noticed in the microcirculation of ZDF rats.

The exercise and low body weight play some roles upon the reducing the calpain expressions or elevating the calpains degradation in the skeletal muscle of NIDDM rats.

Even though a variety of treatments are available for NIDDM treatments and these treatments are not easily available to several economically poorer peoples of the world. A detailed knowledge of the genetic risk factors involving the NIDDM will lead to the development of more specific and cost-effective treatments.

Acknowledgement

It should be acknowledged that this work was partly supported by a grant from the Next-Generation BioGreen 21 Program (No. PJ008191) and a research fund for FTA issues (No. PJ907055), Rural Development Administration, Republic of Korea.

References

[1] Koohmaraie, M. and Geesink, G. H. (2006). Contribution of postmortem muscle biochemistry to the delivery of consistent meat quality with particular focus on the calpain system. *Meat Sci., 74,* 34-43.

[2] Goll, D. E., Thompson, V.F., Li, H. Q., Wei, W. and Cong, J. Y. (2003).The calpain system. *Phy. Rev., 83*, 731-801.

[3] Sorimachi, H., Toyama-Sorimachi, N., Saido, T.C., Kawasaki, H., Sugita, H., Miyasaka, M., Arahata, K., Ishiura, S. and Suzuki, K. J. (1993). Muscle-specific calpain, p94, is degraded by autolysis immediately after translation, resulting in disappearance from muscle. *Biol. Chem., 268*, 10593-10605.

[4] Wendt, A., Thompson, V. F. and Goll, D. E. (2004). Interaction of calpastatin with calpain: A review. *Biol. Chem., 385*, 465-472.

[5] Dayton, W. R., Goll, D. E., Stromer, M. H., Reville, W. J., Zeece, M. G. and Robson, R. M. Some properties of a Ca2+-activated protease that may be involved in myofibrillar protein turnover. In Reich E, Rifkin DB and Shaw E (eds) Cold Spring Harbor Laboratory, Cold Spring Harbor, NY; 1975; pp. 551–557.

[6] Barnoy, S., Glaser, T. and Kosower, N. S. (1997). Calpain and calpastatin in myoblast differentiation and fusion effects of inhibitors. *Biochim. Biophys. Acta, 1358*, 181–188.

[7] Ohno, S., Emori, Y., Imajoh, S., Kawasaki, H., Kisaragi, M. and Suzuki, K. (1984). Evolutionary origin of a calcium-dependent protease by fusion of genes for a thiol protease and a calcium-binding protein. *Nature, 312 (5994),* 566-70.

[8] Glass, J. D., Culver, D. G., Levey, A. I. and Nash, J. (2002). Very early activation of m-calpain in peripheral nerve during Wallerian degeneration. *Neurol. Sci, 196 (1-2),* 9-20.

[9] Cong, J. Y., Gol, D. E., Peterson, A. M. and Kapprell, H. P. (1989). The role of autolysis in activity of the Ca^{2+}-dependent proteinases (μ-calpain and m-calpain). *J. Biol. Chem., 264,* 10096-10103.

[10] Beckmann, J. S. and Spencer, M. (2008). Calpain 3, the “gatekeeper” of proper sarcomere assembly, turnover and maintenance. *Neuromusc. Disord., 18,* 913-21.

[11] Goll, D. E., Kleese, W. C., Okitani, A., Kumamoto, T., Cong, J. and Kapprell, H. P. Intracellular Calcium-dependent Proteolysis. In Mellgren, R.L and Murachi, T. Boca Raton, FL; 1990; CRC. pp. 3–24.

[12] Ono, Y., Kakinuma, K., Torii, F., Irie, A., Nakagawa, K., Labeit, S., Abe, K., Suzuki, K. and Sorimachi, H. (2004). Possible regulation of the conventional calpain system by skeletal muscle-specific calpain, p94/calpain 3. *J. Biol. Chem., 279*, 2761-71.

[13] Dayton, W. R., Schollmeyer, J. V., Lepley, R. A. and Cortes, L. R. (1981). A calcium-activated protease possibly involved in myofibrillar protein turnover. Isolation of a low-calcium-requiring form of the protease. *Biochim. Biophys. Acta*, 659, 48–61.

[14] Mellgren, R. L. (1980). Canine cardiac calcium-dependent proteases: resolution of two forms with different requirements for calcium. *FEBS Lett., 109*, 129–133.

[15] Aoki, K., Imajoh, S., Ohno, S., Emori, Y., Koike, M., Kosaki, G. and Suzuki, K. (1986). Complete amino acid sequence of the large subunit of the low-$Ca2^{+}$-requiring form of human Ca^{2+}-activated neutral protease (muCANP) deduced from its cDNA sequence. *FEBS Lett., 205*, 313–317.

[16] Imajoh, S., Aoki, K., Ohno, S., Emori, Y., Kawasaki, H.,Sugihara, H. and Suzuki, K. (1988). Molecular cloning of the cDNA for the large subunit of the high-$Ca2^{+}$-requiring form of human Ca^{2+}-activated neutral protease. *Biochemistry, 27*, 8122–8128.

[17] Hosfield, C.M., Elce, J. S., Davies, P. L. and Jia, Z. (1999). Crystal structure of calpain reveals the structural basis for Ca^{2+}-dependent protease activity and a novel mode of enzyme activation. *EMBO J., 18,* 6880–6889.

[18] Strobl, S., Fernandez-Catalan, C., Braun, M., Huber, R., Masumoto, H., Nakagawa, K., Irie, A., Sorimachi, H., Bourenkow, G., Bartunik, H., Suzuki, K. and Bode, W. (2000). The crystal structure of calcium-free human m-calpain suggests an electrostatic switch mechanism for activation by calcium. *Proc. Natl. Acad. Sci. USA, 97,* 588–592.

[19] Minami, Y., Emori, Y., Kawasaki, H. and Suzuki, K. (1987). E-F hand structure-domain of calcium-activated neutral protease (CANP) can bind Ca^{2+} ions. *J. Biochem., 101,* 889–895.

[20] Imajoh, S., Kawasaki, H. and Suzuki, K. (1987). The COOH-terminal E-F hand structure of calcium-activated neutral protease (CANP) is important for the association of subunits and resulting proteolytic activity. *J. Biochem., 101,* 447–452.

[21] Suzuki, K., Tsuji, S., Kubota, S., Kimura, Y. and Imahori, K. (1981). Limited autolysis of Ca^{2+}-activated neutral protease (CANP) changes its sensitivity to Ca^{2+} ions. *J. Biochem., 90,* 275–278.

[22] Xu, L. and Deng, X. (2006). Protein kinase Ciota promotes nicotine-induced migration and invasion of cancer cells via phosphorylation of micro- and m-calpains. *J. Biol. Chem., 281*, 4457–4466.

[23] Glading, A., Uberall, F., Keyse, S.M., Lauffenburger, D. A. and Wells, A. (2001). Membrane proximal ERK signaling is required for M-calpain activation downstream of epidermal growth factor receptor signaling. *J. Biol. Chem., 276*, 23341–23348.

[24] Leloup, L., Shao, H., Bae, Y.H., Deasy, B., Stolz, D., Roy, P. and Wells, A. (2010). m-Calpain activation is regulated by its membrane localization and by its binding to phosphatidylinositol 4, 5-bisphosphate. *J. Biol. Chem., 285*, 33549–33566.

[25] Shao, H., Chou, J., Baty, C. J., Burke, N. A., Watkins, S.C., Stolz, D. B. and Wells, A. (2006). Spatial localization of m-calpain to the plasma membrane by phosphoinositide biphosphate binding during epidermal growth factor receptor-mediated activation. *Mol. Cell. Biol., 26*, 5481–5496.

[26] Garcia, M., Bondada, V. and Geddes, J. W. (2005). Mitochondrial localization of mu-calpain. *Biochem. Biophys. Res. Commun., 338*, 1241–1247.

[27] Badugu, R., Garcia, M., Bondada, V., Joshi, A. and Geddes, J. W. N. (2008). Terminus of calpain 1 is a mitochondrial targeting sequence. *J. Biol. Chem., 283*, 3409–3417.

[28] Moldoveanu, T., Gehring, K. and Green, D. R. (2008). Concerted multi-pronged attack by calpastatin to occlude the catalytic cleft of heterodimeric calpains. *Nature, 456*, 404–408.

[29] Averna, M., De Tullio, R., Capini, P., Salamino, F., Pontremoli, S. and Melloni, E. (2003). Changes in calpastatin localization and expression during calpain activation: a new mechanism for the regulation of intracellular Ca_{2+}- dependent proteolysis. *Cell. Mol. Life Sci., 60*, 2669–2678.

[30] Yang, Y. T. and McElligot, M. A. (1989). Multiple actions of the b-adrenergic agonists on skeletal muscle and adipose tissue. *Biochem. J., 261*, 1–10.

[31] Forsberg, N. E., Ilian, M. A., AlBar, A., Checke, P. R. and Wehr, N. B. (1989). Effects of cimaterol on rabbit growth and myofibrillar protein degradation and on calcium-dependent proteinase and calpastatin activities in skeletal muscle. *J. Anim. Sci., 67*, 3313-3321.

[32] Higgins, J.A., Lasslet, Y., Bardsley, R.G. and Buttery, P. J. (1988). The relation between dietary restriction or clenbuterol (a selective b2 agonist) treatment on skeletal muscle growth and calpain proteinase (EC3.4.22.17) and calpastatin activities in lambs. *Br. J. Nutr., 60*, 645-652.

[33] Kretchmar, D.H., Hathaway, M.R., Epley, R.J. and Dayton, W. R. (1989). In vivo effect of a b-adrenergic agonist on activity of calcium-dependent proteinases, their specific inhibitor, and cathepsins B and H in skeletal muscle. *Arch. Biochem. Biophys., 275*, 228–235.

[34] Koohmaraie, M. and Shackelford, S. D. (1991). Effect of calcium chloride infusion on the tenderness of lambs fed a b-adrenergic agonist. *J. Anim. Sci., 69*, 1760-1772.

[35] Cockett, N. F., Jackson, S. P., Shay, T. L., Nielsen, D., Moore, S. S., Stelle, M. R., Barendse, W., Green, R. D. and Georges, M. (1994). Chromosomal localization of the callipyge gene in sheep (Ovis aries) using bovine DNA markers. *Proc. Natl. Acad. Sci., 91*, 3019-3023.

[36] Koohmaraie, M., Shackelford, S.D., Wheeler, T.L., Lonergan, S.M. and Doumit, M. E. (1995). A muscle hypertrophy condition in lamb (callipyge): characterization of effects on muscle growth and meat quality traits. *J. Anim. Sci., 73*, 3596-3607.

[37] Cottin, P., Brutis, J. J., Poussard, S., Elamrani, N., Broncard, S. and Ducastaing, A. (1994). Ca2+-dependent proteinases (calpains) and muscle cell differentiation. *Biochim. Biophys. Acta, 1223*, 170-178.

[38] Murphy, R.M., Verburg, E. and Lamb G. D. (2006). Ca2+ activation of diffusible and bound pools of m-calpain in rat skeletal muscle. *J. Physiol., 576*, 595-612.

[39] Gailly, P., De Backer, F., Van Schoor, M. and Gillis, J. M. (2007). In situ measurements of calpain activity in isolated muscle fibres from normal and dystrophin-lacking mdx mice. *J. Physiol., 582*, 1261-75.

[40] Clark, W. A. (1993). Evidence for posttranslational kinetic compartmentation of protein turnover pools in isolated adult cardiac myocytes. *J. Biol. Chem., 268*, 20243-20251.

[41] Tan, F. C., Goll, D. E. and Otsuka, Y. (1988). Some properties of the millimolar Ca^{2+}-dependent proteinase from bovine cardiac muscle. *J. Mol. Cell. Card., 20*, 983-997.

[42] Van der Westhuyzen, D. R., Matsumoto, K. and Etlinger, J. D. (1981). Easily releasable myofilaments from skeletal and cardiac muscle maintained in vitro. Role in myofibrillar assembly and turnover. *J. Biol. Chem., 256*, 11791-11797.

[43] Dahlmannn, B., Rutschmann, M. and Reinauer, H. (1986). Effect of starvation or treatment with corticosterone on the amount of easily releasable myofilaments in rat skeletal muscle. *Biochem. J., 234*, 659-664.

[44] Reville, W.J., Murphy, B.A., Ahern, S. and Zeece, M. G. (1994). Easily releasable myofilaments in postmortem bovine muscle. *Sciences des Aliments, 14*, 431-444.

[45] Ohtsuka, H., Yajima, H., Maruyama, K. and Kimura, S. (1997). The N-terminal Z-repeat 5 of connectin/titin binds to the C-terminal end of a-actinin. *Biophys. Res. Commun., 235*, 1-3.

[46] Lowell, B.B., Ruderman, N.B. and Goodman M. N. (1986). Evidence that lysosomes are not involved in the degradation of myofibrillar proteins in rat skeletal muscle. *Biochem. J., 234*, 237-240.

[47] Goll, D. E., Kleese, W. C. and Szpacenko, A. (1989). In Campion, D.R., Hausman, G.J., Martin, R.J. (eds). Animal growth regulation. Plenum Publishing Corp., New York, NY. pp.141–181.

[48] Solomon, V. and Goldberg, A. L. (1996). Importance of the ATPubiquitin- proteasome pathway in the degradation of soluble and myofibrilar proteins in rabbit muscle extracts. *J. Biol. Chem., 271*, 26690-26697.

[49] Boehm, M. L., Kendall, T. L., Thompson, V. F. and Goll, D. E. (1998). Changes in the calpains and calpastatin during postmortem storage of bovine muscle. *J. Anim. Sci., 76*, 2415-2434.

[50] Purintrapiban, J., Wang, M.C. and Forsberg, N. E. (2003). Degradation of sarcomeric and cytoskeletal proteins in cultured skeletal muscle cells. *Comp. Biochem. Physiol. B Biochem. Mol. Biol., 136(3)*, 393-401.

[51] Otani, K., Han, D.H., Ford, E.L., Garcia-Roves, P.M., Ye, H., Horikawa, Y., Bell, G.I., Holloszy, J.O. and Polonsky, K. S. (2004). Calpain system regulates muscle mass and glucose transporter GLUT4 turnover. *J. Biol. Chem., 279*, 20915-20920.

[52] Saido, T. C., Sorimachi, H. and Suzuki, K. (1994). Calpain: new perspectives in molecular diversity and physiological-pathological involvement. *FASEB J., 8*, 814-22.

[53] Verburg, E., Murphy, R. M., Richard, I. and Lamb, G. D. (2009). Involvement of calpains in Ca2+-induced disruption of excitation-contraction coupling in mammalian skeletal muscle fibers. *Am. J. Physiol., 296*, C1115-22.

[54] Mellgren, R. L., Zhang, W., Miyake, K. and McNeil P. L. (2007). Calpain is required for the rapid, calcium-dependent repair of wounded plasma membrane. *J. Biol.Chem., 282*, 2567-75.

[55] Kapprell, H.P. and Goll, D. E. (2006). Effect of Ca^{2+} on binding of the calpains to calpastatin. *J. Biol. Chem., 264*, 17888-96.

[56] Ruegg, U. T., Nicolas-Métral, V., Challet, C., Bernard-Hélary, K., Dorchies, O. M., Wagner, S., Buetler, T. M. (2002). Pharmacological control of cellular calcium handling in dystrophic skeletal muscle. *Neuromuscul. Disord., 12(1),* S155-S161.

[57] Isabelle, C.F. and Alexandre, B. (2006). Utrophin is a calpain substrate in muscle cells. *Muscle Nerve, 33*, 753-759.

[58] Fougerousse, F., Gonin, P., Durand, M., Richard, I. and Raymackers, J. M. (2003). Humanmouse differences in the embryonic expression patterns of developmental control genes and disease genes. *Muscle Nerve, 27*, 616-23.

[59] Haravuori, H., Vihola, A., Straub, V., Auranen, M., Richard, I., Marchand, S., Voit, T., Labeit, S., Somer, H., Peltonen, L., Beckmann, J. S. and Udd, B. (2001). Secondary calpain3 deficiency in 2q-linked muscular dystrophy: titin is the candidate gene. *Neurology, 56*, 869-77.

[60] Baghdiguian, S., Richard, I., Martin, M., Coopman, P., Beckmann, J. S., Mangeat, P. and Lefranc, G. (2001). Pathophysiology of limb girdle muscular dystrophy type 2A: hypothesis and new insights into the IkBa/NF-kB survival pathway in skeletal muscle. *J. Mol. Med., 79*, 254-61.

[61] Rey, M.A., and P. L. Davies. 2002. The protease core of the muscle-specific calpain, p94, undergoes Ca2+-dependent intramolecular autolysis. *FEBS Lett., 532*, 401-6.

[62] Hayash, C., Ono, Y., Doi, N., Kitamura, F., Tagami, M., Mineki, R., Arai, T., Taguchi, H., Yanagida, M., Hirner, S., Labeit, D., Labeit, S. and Sorimachi, H. (2008). Multiple molecular interactions implicate the connectin/titin N2A region as a modulating scaffold for p94/calpain 3 activity in skeletal muscle. *J. Biol. Chem., 283*, 14801-14.

[63] Lynch, G.S., Fary, C. J. and Williams, D. A. (1997). Quantitative measurement of resting skeletal muscle [Ca2+]i following acute and long-term downhill running exercise in mice. *Cell Calcium, 22*, 373-83.

[64] Richard, I., Broux, O. and Allamand, V. (1995) Mutations in the proteolytic enzyme calpain 3 cause limb-girdle muscular dystrophy type 2A. *Cell, 81 (1),* 27–40.

[65] Ono, Y., Shimada, H. and Sorimachi, H. (1998) Functional defects of a muscle-specific calpain, p94, caused by mutations associated with limb-girdle muscular dystrophy type 2A. *J. Biol. Chem., 273 (27),* 17073–8.

[66] Biswas, S., Harris, F., Dennison, S., Singh, J. and Phoenix, D. A. (2004). Calpains: targets of cataract prevention? *Trends Mol. Med., 10,* 78-84.

[67] Liu, J., Liu, M. C., Wang, K. K. (2008). Calpain in the CNS: from synaptic function to neurotoxicity. *Sci. Signal, 1 (14),* 1.

[68] Castillo, M. R. and Babson, J. R. (1998). Ca^{2+}-dependent mechanisms of cell injury in cultured cortical neurons. *Neuroscience, 86 (4),* 1133–44.

[69] Lenzlinger, P. M., Saatman, K. E., Raghupathi, R. and Mcintosh, T. K. (2000). Chapter 1: Overview of basic mechanisms underlying neuropathological consequences of head trauma. In Newcomb JK, Miller LS, Hayes RL. Head trauma: basic, preclinical, and clinical directions. New York: Wiley-Liss. ISBN 0-471-36015-5.

[70] Iwata, A., Stys, P. K. and Wolf, J. A. (2004). Traumatic axonal injury induces proteolytic cleavage of the voltage-gated sodium channels modulated by tetrodotoxin and protease inhibitors. *J. Neurosci., 24 (19),* 4605–13.

[71] Turner, M. D., Fulcher, F. K., Jones, C. V., Smith, B. T. and Aganna, E. (2007). Calpain facilitates actin reorganization during glucose-stimulated insulin secretion. *Biochem. Biophys. Res. Commun., 352*, 650–655.

[72] Horikawa, Y., Oda, N., Cox, N. J., Li, X., Orho-Melander, M., Hara, M. and Hinokio, Y. (2000). Genetic variation in the gene encoding calpain-10 is associated with type 2 diabetes mellitus. *Nat. Genet., 26*,163–175.

[73] Baier, L. J., Permana, P. A., Yang, X., Pratley, R. E., Hanson, R. L., Shen, G. Q., Mott, D., Knowler, W. C., Cox, N. J., Horikawa, Y., Oda, N., Bell, G. I. and Bogardus, C. (2000) A calpain-10 gene polymorphism is associated with reduced muscle mRNA levels and insulin resistance. *J. Clin. Invest., 106*, R69–R73.

[74] Ezzidi, A., Turki, A., Messaoudi, S., Chaieb, M., Kacem, M., Al-Khateeb, G. M., Mahjoub, T., Almawi, W.Y. and Mtiraoui, N. (1987). Common polymorphisms of calpain-10 and the risk of Type 2 Diabetes in a Tunisian Arab population: a case-control study. *BMC medical genetics, 11*, ISSN: 1471-2350.

[75] Bodhini, D., Radha, V., Ghosh, S., Sanapala, K. R., Majumder, P. P., Satyanarayana Rao, M. R. and Mohan, V. (2011). Association of calpain 10 gene polymorphisms with type 2 diabetes mellitus in Southern Indians. *Metabolism Clinical and Experimental, 60*, 681–688.

[76] Mazen, M. and Zaharna, E. (2007). Calpain 10 Gene Polymorphism in Type 2 Diabetes Mellitus Patients in Gaza Strip. Thesis. 1428. The Islamic University – Gaza.

[77] Hani, E. H., Stoffers, D. A., Chèvre, J. C., Durand, E., Stanojevic, V., Dina, C., Habener, J. F. and Froguel, P. (1999). Defective mutations in the insulin promoter factor-1 (IPF-1) gene in late-onset type 2 diabetes mellitus. *J. Clin. Invest., 104*, R41–R48.

[78] Macfarlane, W. M., Frayling, T. M., Ellard, S., Evans, J. C., Allen, L. I., Bulman, M. P., Ayres, S., Shepherd, M., Clark, P., Millward, A., Demaine, A., Wilkin, T., Docherty, K. and Hattersley, A. T (1999). Missense mutations in the insulin promoter factor-1 gene predispose to type 2 diabetes. *J. Clin. Invest., 104*, R33–R39.

[79] Walder, K., McMillan, J., Lapsys, N., Kriketos, A., Trevaskis, J., Civitarese, A., Southon, A., Zimmet, P. and Collier, G. (2002). Calpain 3 gene expression in skeletal muscle is associated with body fat content and measures of insulin resistance. *Int. J. Obes. Relat. Metab. Disord., 26*, 442–449.

[80] Tsuchiya T., Schwarz. P.E., Bosque-Plata. L.D., Geoffrey Hayes. M. and Dina, C. (2006). Association of the calpain-10 gene with type 2 diabetes in Europeans: results of pooled and meta-analyses. *Mol. Genet. Metab., 89*, 174–184.

[81] Cox, N. J., Hayes. M. G., Roe, C.A., Tsuchiya, T. and Bell, G. I. (2004). Linkage of calpain 10 to type 2 diabetes: the biological rationale. *Diabetes, 53,* 1S19–25.

[82] Rasmussen, S. K., Urhammer, S. A., Berglund, L., Jensen, J. N. and Hansen, L. (2002). Variants within the calpain-10 gene on chromosome 2q37 (NIDDM1) and relationships to type 2 diabetes, insulin resistance, and impaired acute insulin secretion among Scandinavian Caucasians. *Diabetes, 51*, 3561–3567.

[83] Evans, J. C., Frayling, T. M., Cassell, P. G., Saker, P. J. and Hitman, G. A. (2001). Studies of association between the gene for calpain-10 and type 2 diabetes mellitus in the United Kingdom. *Am. J. Hum. Genet., 69*, 544–552.

[84] Sreenan, S. K., Zhou, Y. P., Otani, K., Hansen, P. A. and Currie, K. P. (2001). Calpains play a role in insulin secretion and action. *Diabetes, 50*, 2013–2020.

[85] Carlsson, E., Poulsen, P., Storgaard, H., Almgren, P. and Ling, C. (2005). Genetic and nongenetic regulation of CAPN10 mRNA expression in skeletal muscle. *Diabetes, 54,* 3015–3020.

[86] Orho-Melander, M., Klannemark, M., Svensson, M. K., Ridderstrale, M. and Lindgren, C. M. (2002). Variants in the calpain-10 gene predispose to insulin resistance and elevated free fatty acid levels. *Diabetes, 51*, 2658–2664.

[87] Zhou, Y. P., Sreenan, S., Pan, C. Y., Currie, K. P. and Bindokas, V. P. (2003). A 48-hour exposure of pancreatic islets to calpain inhibitors impairs mitochondrial fuel metabolism and the exocytosis of insulin. *Metabolism, 52,* 528–534.

[88] Parnaud, G., Hammar, E., Rouiller, D. G. and Bosco, D. (2005). Inhibition of calpain blocks pancreatic beta-cell spreading and insulin secretion. *Am. J. Physiol. Endocrinol. Metab., 289*, E313–321.

[89] Marshall, C., Hitman, G. A., Partridge, C. J., Clark, A. and Ma, H. (2005). Evidence that an isoform of calpain-10 is a regulator of exocytosis in pancreatic beta-cells. *Mol. Endocrinol., 19*, 213–224.

[90] Otani, K., Han, D. H., Ford, E. L., Garcia-Roves, P. M., Ye, H., Horikawa, Y., Bell, G.I., Holloszy, J. O. and Polonsky, K. S. (2004). Calpain system regulates muscle mass and glucose transporter GLUT4 turnover. *J. Biol. Chem., 279*, 20915-20.

[91] Stalker, T. J., Gong, Y. and Scalia, R. (2005). The calcium-dependent protease calpain causes endothelial dysfunction in type 2 diabetes. *Diabetes, 54*, 1132-40.

[92] Carlsson, E., Poulsen, P., Storgaard, H., Almgren, P., Ling, C., Jensen, C. B., Madsbad, S., Groop, L., Vaag, A. and Ridderstrale, M. (2005). Genetic and nongenetic regulation of CAPN10 mRNA expression in skeletal muscle. *Diabetes, 54*, 3015-20.

[93] Hsieh, Y. Y., Chang, C. C., Hsu, K. H., Tsai, F. J., Chen, C. P. and Tsai, H. D. (2008). Effect of exercise training on calpain systems in lean and obese Zucker rats. *Int. J. Biol. Sci., 4(5)*, 300-308.

[94] Griffin, M. E., Marcucci, M. J., Cline, G. W., Bell, K., Barucci, N., Lee, D., Goodyear, L. J., Kragen, E. W., White, M. F. and Shulman, G. I. (1999). Free fatty acid-induced insulin resistance is associated with activation of protein kinase C theta and alterations in the insulin signaling cascade. *Diabetes, 48*, 1270-1274.

[95] Itani, S., Zhou, Q., Pories, W. J., MacDonald, K. G. and Dohm, G. L. (2000). Involvement of protein kinase C in human skeletal muscle resistance and obesity. *Diabetes, 49*, 1353-1358.

[96] Suzuki, K., Hata, S., Kawabata, Y. and Sorimachi, H. (2004). Structure, Activation, and Biology of Calpain. Diabetes, *53(1),* S12-S18.

In: Enzymes and Enzyme Activity
Editor: Eva M. Lashinski

ISBN: 978-1-62417-696-8

Chapter IV

Calpain as Indicator of Organophosphorus-Induced Delayed Neuropathy

Guilherme Luz Emerick** and Lais Silva Fernandes
Faculdade de Ciências Farmacêuticas de Ribeirão Preto-USP,
Ribeirão Preto, S. P., Brazil

Abstract

The intracellular proteases have several functions in regulating various cellular systems. Among them, proteases that are activated by calcium (i.e. calpain), which is a second messenger in cell regulation. Proteases duties consist mainly in the degradation of cytoskeletal proteins, modification of steroid receptors and activation of protein kinases. It has been established that there are at least two types of calpain. One type requires calcium concentrations in the order of mM for their activity, while the other type requires calcium concentrations in the order of μM. Organophosphorus (OPs) compounds, which include pesticides, industrial chemicals and some medicines, are the principal class of

* Corresponding Author: Departamento de Análises Clínicas, Toxicológicas e Bromatológicas, Faculdade de Ciências Farmacêuticas de Ribeirão Preto-USP, Avenida do Café, s/n, Monte Alegre, 14040-903, Ribeirão Preto, S.P., Brazil. Phone: +55 16 3602-0656. Fax: +55 16 3602-4725. E-mail address: *glemerick@yahoo.com.br*.

insecticides and chemical warfare nerve agents. They have become the most widely used insecticides in the world since the 1970s. Clinically, the symptoms of OP poisoning are divided into four categories: acute cholinergic syndrome, intermediate syndrome, chronic organophosphorus-induced neuropsychiatric disorders (COPIND) and organophosphorus-induced delayed neuropathy (OPIDN). OPIDN has been observed in more than 70,000 human cases in the last 75 years and it has been known since the 1930s when tuberculosis patients were treated with phosphocreosote, an uncharacterised mixture of esters derived from phosphoric acid and coal-tar phenols. In 1930, thousands of paralysis cases occurred in the United States due to the consumption of Jamaica ginger extract contaminated with tri-ortho-cresyl phosphate (TOCP). Later, similar epidemics were reported in Morocco, Netherlands, Yugoslavia, France, South Africa, Sri Lanka and India. OPIDN is a neurodegenerative disorder characterised by ataxia progressing to paralysis with a concomitant central and peripheral distal axonopathy, and this neuropathology occurs similar to Wallerian-type degeneration. Initially, neuropathy target esterase (NTE) phosphorylation is involved and then an imbalance in free calcium homeostasis occurs, which leads to the activation of calpain. This sequence of events promotes protein digestion in the terminal axon. Calpain activation has been used to identify Wallerian-type axonal degeneration and this biomarker has been related with the appearance of lesions in long tracts of central and peripheral nervous system. These lesions have caused signs and symptoms that include tingling of the hands and feet. This tingling is followed by sensory loss, progressive muscle weakness and flaccidity of the distal skeletal muscles of the lower and upper extremities and ataxia, which appear about 8-14 days after poisoning. Therefore, calpain evaluation became an excellent tool to indicate axonopathies that affects people around the word.

Keywords: Calpain, Calcium, Organophosphate Pesticides, Organo-phosphorus-induced Delayed Neuropathy

Introduction

The widespread use of pesticides in agriculture in order to increase crop production has been generating a series of toxicological problems. The World Health Organization (WHO), reports about three million cases per year of acute poisoning by pesticides, resulting in 220 thousand deaths, of which 70% occur in developing countries. A survey by the Pan American Health

Organization, in 12 countries of the Latin American and Caribbean, showed that chemical poisoning, especially by lead and pesticides, represents 15% of all occupational diseases reported [1]. Valdez [2] stated that the indiscriminate use of these products has resulted in a serious public health problem and that the quantification of their effects (morbidity and mortality), whether chronic, acute or lethal, is inaccurate due to the large number of underreporting, the lack of specificity of clinical effects and poor knowledge of the health staff about the action of these chemicals and their consequences in intoxicated individuals. This question was reiterated by Pires [3] who showed that the lack of legislation and control of proper use of these products are the main causes of the occurrence of cases of pesticide poisoning. In Brazil, between 1997 and 2001 there were 26,164 cases of poisoning caused by pesticides in the field, representing 7% of all poisonings reported in this period. The main trouble pesticides with regard to acute poisoning globally have been organochlorines (OC), organophosphorus (OP), carbamates and paraquat. Today, most OC insecticides are banned or severely restricted, including in developing countries, thanks to environmentalists who successfully lobbied that their persistence and bioaccumulation implied unacceptable long-term environmental concerns. During the 1960s and 70s, OPs and subsequently carbamate insecticides were introduced in a large scale on the world market. These cholinesterase's inhibitors pesticides, with more rapid breakdown than OCs, were assumed to be less environmentally damaging but with acute toxicity sharper [4]. The OPs are the most commonly used in suicide attempts and led to the greater number of deaths, either by voluntary or involuntary exposure. The OPs monocrotophos and methamidophos were responsible for most of the records of poisonings and suicide attempts, representing an average of 46% and 29% respectively, of the occurrences reported to insecticides in Brazil [3]. Clinically, the symptoms of OP poisoning are divided into four categories: acute cholinergic syndrome, intermediate syndrome (IS), chronic organophosphorus-induced neuropsychiatric disorders (COPIND) and organophosphorus-induced delayed neuropathy (OPIDN) [5]. OPIDN has been observed in more than 70,000 human cases in the last 75 years [6], and it has been known since the 1930s when tuberculosis patients were treated with phosphocreosote, an uncharacterized mixture of esters derived from phosphoric acid and coal-tar phenols [7]. In 1930, thousands of paralysis cases occurred in the United States due to the consumption of Jamaica ginger extract contaminated with tri-ortho-cresyl phosphate (TOCP). Later, similar epidemics were reported in Morocco, Netherlands, Yugoslavia, France, South Africa, Sri Lanka and India [8]. Therefore, it has been reported a large number of poisoning cases in humans, accidental or suicidal, for OPs. Thus, it becomes necessary a better understanding of the mechanism of toxic action of these compounds to prevent the occurrence of serious and irreversible effects.

Role of Calpain in a Normal Cell

Calpain is a calcium-dependent cysteine protease, found in most species studied so far. The first description of calpain was made in 1964 by Guroff [9], and since then, great progress has been made in elucidating the identity, structure, activity, location, physiological and pathological functions. The human genome has 14 distinct genes for calpains [10, 11].

The calpains generally are classified in tissue-specific (also called atypical) and ubiquitous (also called conventional or typical). The calpains tissue-specific have functions related to the tissue in which it is expressed predominantly. For example, it has the skeletal muscle-specific calpain (nCL-1, p94 or calpain 3) and the stomach-specific calpain (nCL 9 or calpain 4). Overall, little is known about the tissue-specific calpains [12]. The ubiquitous calpains exist in mammals as two main isoforms, the calpain I and II, which are encoded by different genes and are the two major isoforms found in tissues including brain [10, 13]. The calpains I and II are also known respectively as μ-calpain, which requires calcium concentrations for its optimal activity in the order of micromolar (5-100 μM) and m-calpain, which requires calcium concentrations for its optimal activity in the order of millimolar (0.2 to 0.8 mM) [12, 14]. The μ and m-calpain are heterodimers composed of two subunits, a larger catalytic subunit (80 kDa), distinct for each calpain (called μCL and mCL respectively for μ and m-calpain) and a small subunit regulatory (30 kDa), which is identical in both. The large subunit (80 kDa) is organized into four domains (I, II, III and IV) and the small subunit (30kDa) is organized into two domains (V and VI) [10, 15].

The two isoforms are found as inactive proenzymes predominantly present in the cytosol, where the concentrations of free calcium is around 50-100 nM in the normal cell at rest. [12]. Several cellular events may increase the activity of calpains, such as increasing the concentration of intracellular free calcium, translocation of calpains to membranes, autolysis of pro-calpains, the subunit dissociation of calpain, decreasing levels of calpastatin (the endogenous inhibitor of calpain) and interaction with calpain activator proteins or phospholipids. The presence and the interaction between these elements are crucial for the adjustment of the calpain activity [16].

Two major mechanisms have been proposed for the activation of calpain *in vivo*. The first proposes that an increased in intracellular concentration of cytosolic free calcium triggers autolysis of the N-terminal portions of the two subunits of the proenzyme, causing a change in conformation of the molecule which leads to the separation of the subunits, occurring activation. Another mechanism proposes that an increase in the concentration of intracellular free calcium leads to a proenzyme translocation from cytosol to the plasma membrane, and then it is activated by proteins and phospholipids present in the membrane, occurring activation without an autolysis and in physiological concentrations of calcium [12, 17].

Immunolocalization studies have shown that the isoforms μ and m-calpain are located exclusively in the intracellular space, except in diseased or injured tissues. Its intracellular localization varies widely in different cell types and tissues [10, 11]. The calpain has specificity for a large number of substrates, such as membrane proteins, steroid receptors, signaling molecules and structural proteins [16]. Their activity toward proteins cytoskeletal and membrane skeleton, such as neurofilaments and microtubules-associated proteins. The calpain are involved in the membrane fusion events, changes in cell configuration, including neurite extension and remodeling of the synaptic terminals [10, 18, 19]. Other roles include apoptosis and cell death, cell motility, signal transduction, and fusion of myoblasts [16, 20, 21]. The cleavage of cytosqueletal and membrane attachments are probably the most well-documented role of calpain.

In vitro studies demonstrate that many kinases (ppFAK125, pp60c-src), phosphatases (tyrosine phosphatase, PTP-1B protein), and cytoskeletal proteins (talin, filamin, paxillin, vinculin) that are involved in signal transduction pathways are rapidly cleaved by calpains [11]. Some proteins involved in apoptosis (caspase-3, Bax, Bcl-XL, and Bid) are substrates for calpain. The hidrolysis of these proteins mediated by calpains cause changes in their activity or intracellular localization, leading to apoptosis [16].

The calpastatin protein is an endogenous inhibitor of calpain and it has not been demonstrated that this inhibitor is able to inhibit other proteases. Calpastatin co-exists with calpain in the cytosol and one molecule of this natural inhibitor can react with four molecules of calpain. One step in determining the role of calpain and calpain isoform in the physiology and pathology of the cell has been the lack of good pharmacological inhibitors [16, 22].

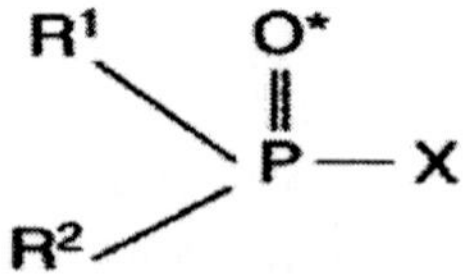

*or S

Figure 1. General chemical structure of an organophosphate pesticide.

Organophosphorus Pesticides

The first reported organophosphorus cholinesterase inhibitor was tetraethyl pyrophosphate (TEPP), which was first synthesized in 1854 by Philip de Clermount [23]. Chemically, the OPs (Figure 1) usually are esters, thiol esters or acid anhydrides derivatives of phosphorus-containing acids and have become the most widely used insecticides in the world from the 70s because of their property of inhibiting acetylcholinesterase (AChE) in insects. In the Figure 1 R1 and R2 are usually simple alkyl or aryl groups both of which may be bonded directly to the phosphorus atom (in phosphinates), or linked via -O-, or -S- (in phosphates), or R1 may be bonded directly and R2 via one of the above groups (phosphonates). In phosphoroamidates, carbon is linked to phosphorus through an amino group. The group X is referred to as the leaving group and can be any one of a wide variety of substituted and/or branched aliphatic, aromatic or heterocyclic groups, linked to phosphorus via a bond of some lability (usually -O- or -S-) [24].

The organophosphate insecticides are absorbed through the skin, respiratory and gastrointestinal tract, and often their absorption is favored by solvents present in the formulation. After absorption, these compounds and their metabolites are rapidly distributed to all tissues. For all compounds that contain a sulfur bound to the phosphorus, a metabolic bioactivation is necessary for their biological activity to be manifest because only compounds with P=O moiety are effective inhibitor of AChE. However, these oxidized analogs can be rapidly hydrolyzed by hydrolases found in the tissues of mammals. The insects are more sensitive to these compounds because they often do not possess these enzymes [25]. As can be seen in Figure 2 is required that occurs oxidation of ethyl parathion to ethyl paraoxon in order to allow the strong inhibition of AChE. Initially, in the chemical interactions between the OP and AChE forms a Michaelis complex enzyme-OP where a specific serine in the protein is phosphorylated with loss of the leaving group "X".

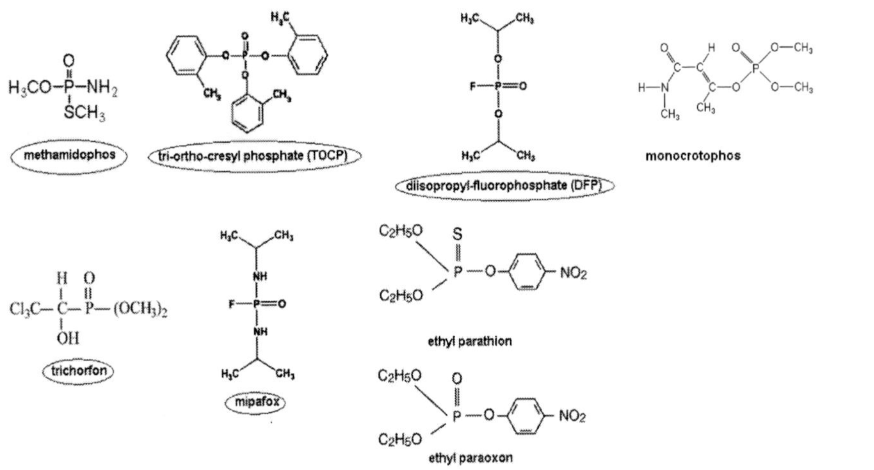

Figure 2. Chemical structures of some OPs. Circulated are inducers of delayed neuropathy.

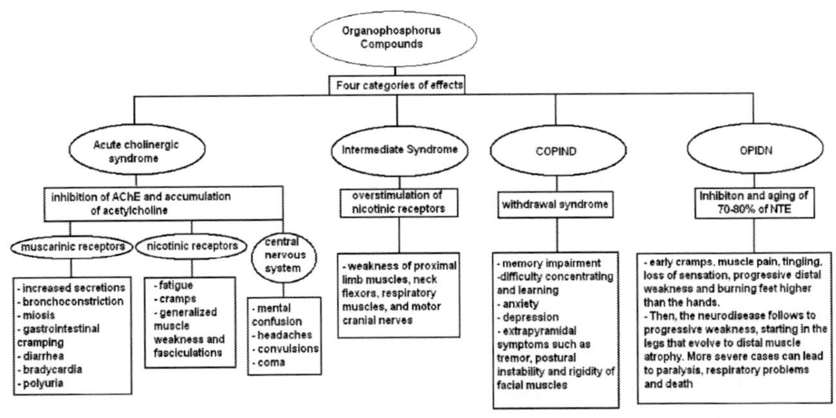

Figure 3. Summarize of main effects of OPs.

This is a progressive reaction leading to formation of a reasonably stable covalent bond between OP compound and enzyme with consequent inhibition of catalytic activity. Reactivation may occur spontaneously but slowly at a rate that is dependent both on the nature of the attached group and on the enzyme protein; the rate can be influenced by added nucleophilic reagents, such as oximes, which may accelerate the reaction and thereby act as antidotes. On the other hand, may occurs a second reaction known as "aging" with the loss of one of the radicals attached to phosphorus via a bond-O- or -S- and the consequent formation of an negative complex enzyme-OP$^-$. This reaction is called aging because this is time-dependent and the enzyme-OP complex is highly stable to the reactions of reactivation. But to place the signs from the inhibition of AChE a simple inhibition is already enough.

The Organophosphorus Intoxication

As a primary target of OPs, AChE is a β-hydrolase that the physiological role is hydrolyze the neurotransmitter acetylcholine in the central and peripheral nervous system. But, some OPs have been shown to produce long-term neurological damage independent of their effects on AChE inhibition [26, 27]. These effects are known as COPIND and OPIDN. Among the delayed effects caused by OPs, the OPIDN has been most studied by several authors [28-33]. There is another effect that can occur between the onset of cholinergic syndrome and chronic effects known as IS. In the next sections, we discuss each effect separately. Figure 3 summarize the main effects of OPs.

Cholinergic Syndrome

Acetylcholine is the transmitter in synapses in the parasympathetic nervous system, in the autonomic ganglia of the sympathetic nervous system and at the motor end-plate. These cholinergic synapses/receptors are divided into muscarinic receptors stimulated by muscarine (parasympathetic to the heart, smooth muscles and secretory glands) and nicotinic receptors stimulated by nicotine (autonomic ganglia and motor end-plates). Under normal conditions, AChE rapidly and efficiently hydrolyzes acetylcholine. Prior to inactivation, acetylcholine molecules interact with postsynaptic cholinergic receptors to alter cellular function, either by alterating ion flux across the postsynaptic cell membrane or through the generation of intracellular second messengers. Elevated synaptic acetylcholine levels lead to persistent stimulation of cholinergic receptors on postsynaptic cells and subsequent alteration of cholinergic receptor-mediated signaling pathways. These cellular changes lead to functional changes at the tissue level [34].

While all cholinesterase inhibitors are thought to elicit toxicity through this general mechanism, the spectrum of cholinergic signs can be different with different inhibitors, the same inhibitor in different species, or the same inhibitor with different routes of exposure. Signs and symptoms of acute poisoning with anticholinesterases agents are predictable from their biochemical mechanism of action and are directly related to the levels of AChE activity [35]. Typical manifestations of cholinergic crisis include, but are not limited to, nausea, vomiting, diarrhea, abdominal cramp, urinary incontinence, miosis, salivation, lacrimation, bronchorrhea, bradycardia, hypotension, fasciculation, muscle paralysis, dizziness, confusion, seizures, coma, and respiratory failure. Death can occur within a very short space of time if life threatening conditions, such as respiratory failure, are not treated promptly and appropriately. The signs due to stimulation of the parasympathetic autonomic nervous system may come from the action on muscarinic or nicotinic receptors. The stimulation of muscarinic receptors leads to increased secretions, bronchoconstriction, miosis, gastrointestinal cramping, diarrhea, bradycardia, and polyuria. Since the effects on nicotinic receptors are: fatigue, cramps, generalized muscle weakness and fasci-culations. The effects on the central nervous system are mental confusion, headaches, convulsions and coma [36]. These symptoms require emergency care. The first 4–6 h are the most critical in acute poisoning with OP pesticides. If there is improvement in symptoms after initial treatment then the patient is very likely to survive if adequate treatment is continued. Clinical diagnosis is relatively simple and is based on medical history, circumstances of exposure, clinical presentation, and laboratory tests. Confirmation of diagnosis can be made by measurement of erythrocyte AChE or plasma cholinesterase. Activities of these enzymes are accepted as biomarkers of exposure and/or toxicity of OP and carbamate. Erythrocyte AChE is identical to the enzyme present in the target synapses and its levels are assumed to reflect the effects in target organs. Thus, erythrocyte AChE is regarded as biomarker of toxicity of these compounds [35]. For treatment it is usually administered atropine which acts as competitive inhibitor of acetylcholine and oximes that act as reactivators of AChE.

Atropine is the first drug used in acute OP intoxication and acts by blocking the effects of excess concentrations of acetylcholine at muscarinic cholinergic synapses. Atropine can relieve the following symptoms of OP poisoning: sweating, salivation, rhinorrhoea, lacrimation, nausea, vomiting and diarrhea, and can help control of bradycardia and circulatory depressions. Atropine does not bind to nicotinic receptors and cannot relieve nicotinic effects in OP poisoning. Atropine is preferable given intravenously, though the intramuscular route is also effective. It also has used in association with atropine oximes that reactivate phosphorylated cholinesterases by displacing the phosphoryl moiety from the enzyme by virtue of their high affinity for the enzyme and their powerful nucleophilicity. However, this chemical reaction occurs only when the phosphorylated AChE has not undergone aging. In addition to oximes, diazepam is also used in the treatment of acute OP poisoning to relieve anxiety in mild cases and to reduce muscle fasciculations and antagonize convulsions in the more severe cases [37].

Intermediate Syndrome

The term IS was first described by Senanayake and Karalliedde in 1987 [38], because it arose in the interval between the end of the acute cholinergic crisis and the onset of OPIDN. According to that report, IS was characterized by weakness of proximal limb muscles, neck flexors, respiratory muscles, and motor cranial nerves, and was attributed to muscle fiber necrosis following acute cholinergic crisis. The IS is seen in about 80% of acute OP poisoning cases and was originally seen following the ingestion of highly lipophilic OP compounds, such as fenthion, but it is not confined to a few distinct compounds [39]. Although the incidence of IS may be high and the syndrome has been considered a major contributing factor of organophosphate-related morbidity and mortality, the pathophysiology that underlies IS remains unclear. The syndrome develops 24 to 96 hours after OP intoxication, during recovery from cholinergic manifestations, or in some cases, when patients are completely recovered from the initial cholinergic crisis and was not related to the incipient delayed neuropathy discussed in a later topic of this chapter.

Senanayake and Karalliedde [38] in their first report of IS suggested that the syndrome might be caused by pathologic changes in the postsynaptic end-plate region of striated muscles because such lesions were described in experimental animals (hens and rats) that developed a similar pattern of paralysis after OP poisoning. On the other hand, Benson and McIntire [40] did not agree and suggested that either different susceptibility of various cholinergic receptors or inadequate oxime therapy played a role in the development of IS. They stated that IS could be explained by the above findings, which were due to a reduction in the number of functioning acetylcholinesterase receptors at the post-junctional membrane of nicotinic receptors or a failure of acetylcholine release. There is no specific treatment for the IS and intervention is exclusively supportive.

Chronic Organophosphorus-Induced Neuropsychiatric Disorders

COPIND can be caused by repeated exposure and low doses. The mechanism of this syndrome is the least known of the chronic effects of OPs, but it is known that this mechanism is independent of inhibition of esterases. The most common clinical signs include memory impairment, difficulty concentrating and learning, anxiety, depression and extrapyramidal symptoms such as tremor, postural instability and rigidity of facial muscles that do not respond to treatment with levodopa. Regarding psychiatric symptoms, subjects heavily exposed to pesticides (40 h/week, 9 months/year) had a significant increase in the frequency of psychiatric disorders, especially depressive neurosis and dysthymic disorder. Among OP commonly used in our population, chlorpyriphos has been reported to induce these syndromes, including anxiety, depression, and transient parkinsonism [5].

COPIND usually persists long and is delay-occurred. Farm workers tested about two years after a pesticide poisoning episode showed significantly lower performance in verbal and visual attention and visual memory. Tan et al. [41] characterized these chronic effects as a withdrawal of OP pesticide after chronic low-level exposure or acute exposure. This biological syndrome is a phenomenon induced by adaptations that can diminish the effects of a drug in response to repeated exposure, while cessation of drug use; the adaptations do not immediately disappear, resulting in the emergence of drug-opposite responses as withdrawal syndrome. This hypothesis can help to explain why COPIND is delay-occurred, usually appears days to months after OP exposure.

Organophosphorus-Induced Delayed Neuropathy

OPIDN is characterized by a distal axonopathy, central and peripheral Wallerian-type degeneration that develops 8-14 days after poisoning by a neuropathic OP. Among the OPs that cause OPIDN include phosphates, phosphonates and phosphoroamidates. As an example of some compounds that have been reported as causing OPIDN have tri-o-cresyl phosphate, methamidophos, mipafox, dichlorvos, trichlorfon and leptophos. The methamidophos (O,S-dimethyl phosphoramidothioate) has been the pesticide most frequently reported as causing OPIDN in humans [42]. For the development of this neuropathy is a need of 70-80% of inhibition of a membrane-bound esterase known as neuropathy target esterase (NTE). The usual symptoms are early cramps, muscle pain, tingling, loss of sensation, progressive distal weakness and burning feet higher than the hands. Then, the neurodisease follows to progressive weakness, starting in the legs that evolve to distal muscle atrophy. More severe cases can lead to paralysis, respiratory problems and death [36]. In Brazil, there are reports of delayed neuropathy induced by OPs involving rural workers. A study at the region of Nova Friburgo, in the state of Rio de Janeiro, found a prevalence of 12.8% of neuropathy cases compared to all the signs of poisoning by pesticides. One of the factors contributing to this high prevalence was excessive use of OPs compounds for combating insects, including the active ingredient methamidophos that accounted for 90% of utilization among all pesticides used in that region [43].

The mechanism to generate OPIDN was first described by Johnson [29] and initially involves neuropathy target esterase (NTE) phosphorylation. Simple inhibition of NTE by OP is not sufficient to result in OPIDN (different from AChE). It is necessary to generate a negative charge on the terminal portion of the phosphate group bonded to the enzyme. This occurs as a result of a second reaction, so called "aging". In this step, the cleavage of one bond in the R-O-P chain, with the loss of R leads to formation of a charged mono-substituted phosphoric acid residue still attached to the protein (serine residue).

This reaction is called aging because it is a slow progressive process and the product is no longer responsive to nucleophilic reactivating agents such as oximes. This complex, (enzyme-organophosphate) is very stable and resistant to breakage or hydrolytic reactivation. The rate of aging of the phosphorylated enzyme is dependent on the nature of the R groups, generally being slowest with methyl groups [26].

Furthermore, it also occurs an imbalance in free calcium homeostasis which leads to the activation of calcium-activated neutral protease (CANP) [30-32]. These events promote protein digestion in the terminal axon which seems to proceed similarly to Wallerian degeneration. There are many variables involved in the biochemical mechanism of neuropathic OPs' toxic action.

Some OPs, such as paraoxon, strongly inhibit acetylcholinesterase (AChE), but they neither inhibit NTE nor cause OPIDN in *in vivo* studies. Others OPs, such as diisopropyl phosphorofluoridate (DFP), mipafox and TOCP, inhibit both AChE and NTE and cause OPIDN [44]. For example, phosphates, phosphonates and phosphoramidate are neuropathics, and sulfonates, carbamates, thiocarbamates, sulfonyl fluorides and phosphinates are not. Some studies report that, with some chemicals (e.g., methamidophos), the (+)-enantiomer causes OPIDN in hen while the (-)-enantiomer does not. This is because the negative form has a higher affinity for AChE, not for NTE [31, 45, 46].

Several animal species are sensitive to neuropathic OPs. These lists include chickens, cats, dogs, horses, oxen, sheep, pigs, monkeys and humans [47]. The adult hen is the animal of choice for studying this neuropathy experimentally, due to its sensitivity and clinical signs are very similar to those seen in humans, moreover, is a domestic animal and easy to use [7, 48].

Neuropathy Target Esterase

As a target of neuropathic OP, NTE was first identified by Johnson [28] and its inhibition and aging is an initial event in the appearance of the delayed neuropathy in hens. It is a membrane-bound enzyme, with a molecular weight of 155 kDa. Even though it is known that this enzyme is involved in the homeostasis of the phosphatidylcholines [49], its physiological role is not yet defined [50]. But, in a recent study Greiner et al. [51] have proposed that the physiological enzymatic activity of NTE may be manifested in maintaining bilayer membrane fluidity in intact cells. Lysophospholipids in healthy mammalian cell membranes constitute between 0.5% and 6% of the total lipid weight. However, at elevated concentrations, these lipids have been linked to a number of disease states that include atherosclerosis, inflammation, and hyperlipidemia. It has been well established that lysophospholipids act as neurotoxic agents when added to cell types such as cerebellar granular neurons and rat spinal cord motor neurons at micromolar concentrations in vitro. Therefore, NTE normal activity is essential for the maintenance of the neuron membrane and consequently for correct functions of the membrane permeability.

According to Glynn [26] NTE may belong to the family of serine hydrolases or belong to a new family of proteins. AChE is by far the best characterized serine esterase. AChE is a member of a large family of proteins with related primary sequences including enzymes which hydrolyze a diverse spectrum of carboxyl esters ranging from acetate esterified with choline (acetylcholine) to a variety of long chain fatty acids esterified with glycerol (triacylglcerol). The serine esterase and protease families are both included in a larger functional grouping of enzymes which all have a reactive nucleophile at their active site together in a catalytic triad with a histidine and an acidic (aspartate or glutamate) residue. On the other hand, the primary sequence of NTE is unrelated to proteins of the AChE family or indeed to any known serine hydrolases; instead NTE is a member of a novel protein family represented in organisms from bacteria to man.

The activity of the NTE can be measured in the central nervous system and peripheral nerves, as well as in the lymphocytes and platelets [52]. Oliveira et al. [53] monitoring the activity of NTE in the central nervous system of hens, found that the maximum inhibition of the enzyme occurred 24 hours after TOCP administration.

Inhibitors of NTE, such as phosphinates, carbamates, and sulfonyl fluorides, which react covalently with the NTE active center, but do not promote the enzyme aging reaction, do not form a terminal group with a negative charge, so the enzyme can be reactivated by endogenous mechanisms.

When some of these NTE inhibitors are given to birds before neuropathic OP, aging of the enzyme-OP complex does not occur rather, there is a protection of the bird, preventing the subsequent OPIDN. However, when one of them is given after neuropathic OP, they normally exert a potentiating action, increasing the OPIDN neuropathic clinical signs; there is a promotion [54].

Wallerian Degeneration

In several neurodegenerative disorders caused by toxic agents, Wallerian degeneration occurs. According to the sequence of events that occur in the distal axon, the axonal degeneration is called Wallerian type. It is characterized by activation of calcium-activated neutral protease (calpains) due to excessive intake of calcium in the cell.

Activation of calpain promotes digestion of the terminal portion of the axon, thus preventing the transmission of nerve impulses to the postsynaptic cell. When the injury occurs by chemical agents in the cell body of the neuron, is called the neuronal degeneration. But when the injury occurs in the axon, it is called axonal degeneration as in the case of OPIDN. Initially, there is a period which the distal stump survives and maintains relatively normal structural, transport and conductions properties.

This duration is proportional to the length of the axonal stump. Terminating the survival period occurs proteolysis that digests the axolemma and axoplasm, leaving only a myelin sheath surrounding a swollen degenerate axon. Although it is established that degeneration of the most terminal portion of the axon occurs first, whether degeneration of the remainder of the stump occurs from the proximal to distal, distal to proximal, or simultaneously along its entire length remains a matter of debate [55].

Although the inhibition and subsequent aging of NTE has been proposed as initiating events in OPIDN, the events that occur between NTE inhibition and the appearance of clinical effects are not completely understood. It has been proposed that NTE inhibition by OPs is associated with an increase in intracellular Ca^{2+} and a decrease in serum Ca^{+2} [56]. Calcium migrates into the cell easily by means of altered calcium channels and activated calpain, which provokes the neurodegenerative process.

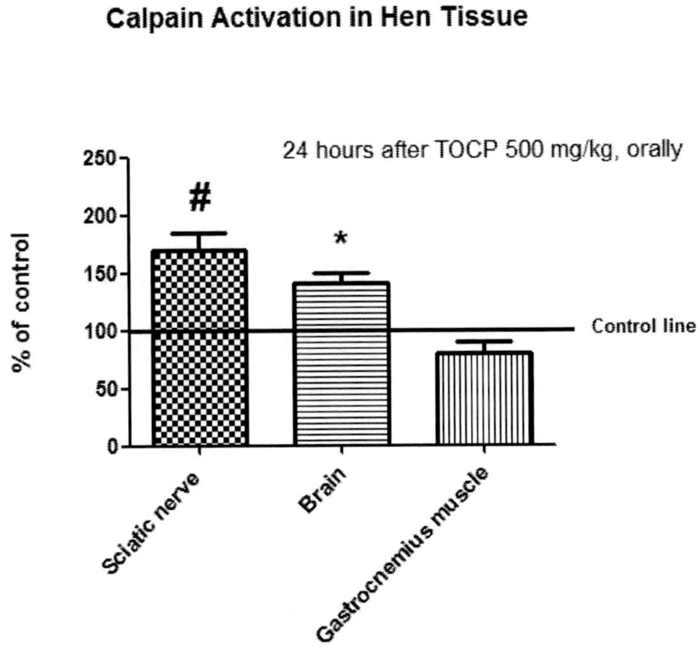

Figure 4. Calpain activity in hen tissues 24 hours after single dose of TOCP (500 mg/kg), orally. The tissues were analyzed as shown in other works [30, 31]. N = 3, # a statistically significance difference compared with the control and brain group ($p < 0.05$, according to ANOVA followed by Tukey's test for multiple comparisons); * a statistically significance difference compared with the control group ($p < 0.05$, according to ANOVA followed by Tukey's test for multiple comparisons). Values are represented as means ± SEM.

This increase in Ca^{2+} may then activate Ca^{2+}/calmodulin-dependent kinase leading to aberrant phosphorylation of cytoskeletal proteins and their accumulation, which is a common feature of many neurodegenerative diseases.

Previous studies have shown that neurodegeneration and the breakdown of cytoskeletal elements are accompanied by increased levels of axonal free calcium [57]. After tri-*ortho*-cresyl phosphate (TOCP) administration, alterations in calcium homeostasis are observed in diverse tissues due to increased calcium membrane permeability [30]. As consequence of this increase in intracellular calcium, calpain activation is noted in nervous tissues but not in muscle tissue proving that this is neuropathology and not muscle pathology (Figure 4). The intracellular proteases have several functions in regulating various cellular systems. Among them, proteases that are activated by calcium, which is a second messenger in cell regulation. Proteases duties consist mainly in the degradation of cytoskeletal proteins, modification of steroid receptors and activation of protein kinases [58].

References

[1] WHO. Public health impact of pesticides used in agriculture. World Health Organization. 1990.

[2] Valdez N. E. C. Monitorização biológica de trabalhadores expostos a inseticida organofosforado neurotoxicante: uma proposta de intervenção. ENSP/FIOCRUZ, Rio de Janeiro. 2005.

[3] Pires D. X., Caldas E. D., Recena M. C. [Pesticide use and suicide in the State of Mato Grosso do Sul, Brazil]. Cadernos de saude publica / Ministerio da Saude, Fundacao Oswaldo Cruz, *Escola Nacional de Saude Publica*. 2005;21(2):598-605. Epub 2005/05/21. Uso de agrotoxicos e suicidios no Estado do Mato Grosso do Sul, Brasil.

[4] Wesseling C., Corriols M., Bravo V. Acute pesticide poisoning and pesticide registration in Central America. *Toxicology and applied pharmacology*. 2005;207(2 Suppl.):697-705. Epub. 2005/09/13.

[5] Salvi R. M., Lara D. R., Ghisolfi E. S., Portela L. V., Dias R. D., Souza D. O. Neuropsychiatric evaluation in subjects chronically exposed to organophosphate pesticides. *Toxicological sciences : an official journal of the Society of Toxicology*. 2003;72(2):267-71. Epub. 2003/03/28.

[6] Johnson M. K., Glynn, P. Neuropathy target esterase. In: Krieger R. I., editor. *Handbook of Pesticide Toxicology*. San Diego: Elsevier; 2001. p. 953-65.

[7] Johnson M. K. Structure-activity relationships for substrates and inhibitors of hen brain neurotoxic esterase. *Biochemical pharmacology*. 1975;24(7):797-805. Epub. 1975/04/01.

[8] Zhao X. L., Zhu Z. P., Zhang T. L., Zhang C. L., Yu L. H., Xie K. Q. Tri-ortho-cresyl phosphate (T. O. C. P.) decreases the levels of cytoskeletal proteins in hen sciatic nerve. *Toxicology letters*. 2004; 152(2): 139-47. Epub. 2004/08/11.

[9] Guroff G. A neutral calcium-activated proteinase from the soluble fraction of rat brain. *J. Biol. Chem*. 1964;239:149–55.

[10] Grynspan F., Griffin, W. B., Mohan, P. S., Shea, T. B., Nixon, R. A. Calpains and Calpastatin in SH-SY5Y Neuroblastoma Cells During Retinoic Acid-Induced Differentiation and Neurite Outgrowth: ComparisonWith the Human Brain Calpain System. *Journal of Neuroscience Research*. 1997;48:181–91.

[11] Goll D. E., Thompson V. F., Li H., Wei W., Cong J. The calpain system. *Physiological reviews*. 2003;83(3):731-801. Epub. 2003/07/05.

[12] Ray S. K., Hogan E. L., Banik N. L. Calpain in the pathophysiology of spinal cord injury: neuroprotection with calpain inhibitors. *Brain research Brain research reviews*. 2003;42(2):169-85. Epub. 2003/05/10.

[13] Banik N. L., Matzelle D. C., Gantt-Wilford G., Osborne A., Hogan E. L. Increased calpain content and progressive degradation of neurofilament protein in spinal cord injury. *Brain research*. 1997;752(1-2):301-6. Epub. 1997/03/28.

[14] Kishimoto A., Kajikawa N., Tabuchi H., Shiota M., Nishizuka Y. Calcium-dependent neural proteases, widespread occurrence of a species of protease active at lower concentrations of calcium. *Journal of biochemistry*. 1981;90(3):889-92. Epub. 1981/09/01.

[15] Sorimachi H., Ono Y. Regulation and physiological roles of the calpain system in muscular disorders. *Cardiovascular research*. 2012. Epub. 2012/05/01.

[16] Liu X., Van Vleet T., Schnellmann R. G. The role of calpain in oncotic cell death. *Annual review of pharmacology and toxicology*. 2004;44:349-70. Epub. 2004/01/28.

[17] Melloni E., Michetti M., Salamino F., Pontremoli S. Molecular and functional properties of a calpain activator protein specific for mu-isoforms. *The Journal of biological chemistry*. 1998;273(21):12827-31. Epub. 1998/05/28.

[18] Song D. K. M. T., Kater S. B., Mykles D. L. Calpain inhibitors block Ca (+2)-induced suppression of neurite outgrowth in isolated hippocampal pyramidal neurons. *J. Neurosci*. 1994;39:474–81.

[19] Shea T. B. C. C. M., Spencer M. J., Beermann M. L., Nixon R. A. Enhancement of neurite outgrowth following calpain inhibition is mediated by proteinase kinace C. *J. Neurochem*. 1995;65:517–27.

[20] Glading A., Lauffenburger D. A., Wells A. Cutting to the chase: calpain proteases in cell motility. *Trends in cell biology*. 2002;12(1):46-54. Epub. 2002/02/21.

[21] Macqueen D. J., Meischke L., Manthri S., Anwar A., Solberg C., Johnston I. A. Characterisation of capn1, capn2-like, capn3 and capn11 genes in Atlantic halibut (Hippoglossus hippoglossus L.): Transcriptional regulation across tissues and in skeletal muscle at distinct nutritional states. *Gene*. 2010;453(1-2):45-58. Epub. 2010/01/23.

[22] Averna M., de Tullio R., Passalacqua M., Salamino F., Pontremoli S., Melloni E. Changes in intracellular calpastatin localization are mediated by reversible phosphorylation. *The Biochemical journal*. 2001;354 (Pt 1):25-30. Epub. 2001/02/15.

[23] Pope C., Karanth S., Liu J. Pharmacology and toxicology of cholinesterase inhibitors: uses and misuses of a common mechanism of action. *Environmental toxicology and pharmacology*. 2005;19(3): 433-46. Epub. 2005/05/01.

[24] Johnson M. K., Jacobsen, D., Meredith, T. J., Eyer, P., Heath, A. J., Ligtenstein, D. A., Marrs, T. C., Szinica, L., Vale, J. A., Haines, J. A. Evaluation of antidotes for poisoning by organophosphorus pesticides. *Emerg. Med*. 2000;12:22-37.

[25] Costa L. G. Toxic effects of pesticides In: Klaassen C. D., editor. *Casarett and Doulls toxicology: the basic science of poisons*. New York: McGraw-Hill, 978-0-07-134721-1; 2007. p. 883-930.

[26] Glynn P. Neural development and neurodegeneration: two faces of neuropathy target esterase. *Progress in neurobiology*. 2000;61(1):61-74. Epub. 2000/04/12.

[27] Glynn P. A. mechanism for organophosphate-induced delayed neuropathy. *Toxicology Letters*. 2006;162:94-7.

[28] Johnson M. K. The delayed neurotoxic effect of some organophosphorus compounds. Identification of the phosphorylation site as an esterase. *The Biochemical journal*. 1969;114(4):711-7. Epub. 1969/10/01.

[29] Johnson M. K. The target site for the initiation of delayed neurotoxicity by organophosphorous esters: biochemical studies and toxicological applications. In: Hodgson J. R. P., R. M., editor. *Rev. Biochem. Toxicol*. 1982. p. 141-212.

[30] Emerick G. L., Peccinini R. G., de Oliveira G. H. Organophosphorus-induced delayed neuropathy: a simple and efficient therapeutic strategy. *Toxicology letters*. 2010;192(2):238-44. Epub. 2009/11/17.

[31] Emerick G. L., DeOliveira G. H., Oliveira R. V., Ehrich M. Comparative in vitro study of the inhibition of human and hen esterases by methamidophos enantiomers. *Toxicology*. 2012;292(2-3):145-50. Epub. 2011/12/27.

[32] Emerick G. L., Deoliveira G. H., Dos Santos A. C., Ehrich M. Mechanisms for consideration for intervention in the development of organophosphorus-induced delayed neuropathy. *Chemico-biological interactions*. 2012. Epub. 2012/07/24.

[33] Emerick G. L., Oliveira R. V., Belaz K. R., Goncalves M., DeOliveira G. H. Semipreparative enantioseparation of methamidophos by HPLC-UV and preliminary in vitro study of butyrylcholinesterase inhibition. *Environmental toxicology and chemistry / S. E. T. A. C.* 2012; 31(2):239-45. Epub. 2011/11/03.

[34] Pope C., Karanth, S., Liu, J. . Pharmacology and toxicology of cholinesterase inhibitors: uses and misuses of a common mechanism of action. *Environmental Toxicology and Pharmacology*. 2005;19:433-46.

[35] Jokanovic M. Medical treatment of acute poisoning with organophosphorus and carbamate pesticides. *Toxicology letters*. 2009; 190(2):107-15. Epub. 2009/08/05.

[36] Ecobichon D. J. Toxic effects of pesticides. In: Klaassen C. D., editor. Casarett and Doull's Toxicology:*The Basic Science of Poisons*. New York: McGraw-Hill, 978-0-07-134721-1; 2001. p. 763-85.

[37] Lotti M. Clinical toxicology of anticholinesterases in humans. In: Academic Press ----, editor. *Handbook of Pesticide Toxicology*. San Diego 2001.

[38] Senanayake N., Karalliedde, L. Neurotoxic effects of organophosphorus insecticides: an intermediate syndrome. *New England Journal of Medicine*. 1987;316:761–3.

[39] Yang C., Deng, J. . Intermediate syndrome following organophosphate insecticide poisoning. *Journal of Chinese Medical Association.* 2007;70: 467-72.

[40] Benson B. J., McIntire, M. . Is the intermediate syndrome in organophosphate poisoning the result of insufficient oxime therapy?, *Journal of Toxicology - Clinical Toxicology*. 1992;30:347–9.

[41] Tan D., Peng, S., Wu, Y., Wang, Y., Lu, C., Yan, C. Chronic organophosphate (OP)-induced neuropsychiatric disorder is a withdrawal syndrome. *Medical Hypothesis*. 2009;72:405-6.

[42] Mcconell R., Téllez, E. D., Cuadra, R., Torres, E., Keifer, M., Almendárez, J., Miranda, J., El-Fawal, H. A. N., Wolff, M., Simpsom, D., and Lundberg, I. Organophosphate neuropathy due to methamidophos: biochemical and neurophysiological markers. *Archives of Toxicology*. 1999;73:296-300.

[43] Araújo A. J., Lima, J. S., Moreira, J. C., Jacob, S. C., Soares, M. O., Monteiro, M. C. M., Amaral, A. M., Kubota, A., Meyer, A., Cosenza, C. A. N., Neves, C., Markowitz, S. Exposição múltipla a agrotóxicos e efeitos à saúde: estudo transversal em amostra de 102 trabalhadores rurais. Ciências e Saúde Coletiva, Nova Friburgo, R. J. 2007;12:115-30.

[44] Hong M. S., Hong, S. J., Barhoumi, R., Robert, C., Burghardt, R. C., Donnelly, K. C., Wild, J. R., Venkatraj, V., Tiffany-Castiglioni, E. Neurotoxicity induced in differentiated SK-N-SH-SY5Y human neuroblastoma cells by organophosphorus compounds. *Toxicology and Applied Pharmacology*. 2003;186:110-8.

[45] Battershill J. M., Edwards, P. M., Johnson, M. K. Toxicological assessment of isomeric pesticides: a strategy for testing of chiral organophosphorus (O. P.) compounds for delayed polineuropathy in a regulatory setting. *Food and Chemical Toxicology*. 2004;42:1279-85.

[46] Bertolazzi M., Caroldi S., Moretto A., Lotti M. Interaction of methamidophos with hen and human acetylcholinesterase and neuropathy target esterase. *Archives of toxicology*. 1991;65(7):580-5. Epub. 1991/01/01.

[47] Johnson M. K., Glynn P. Neuropathy target esterase (N. T. E.) and organophosphorus-induced delayed polyneuropathy (O. P. I. D. P.): recent advances. *Toxicology letters*. 1995;82-83:459-63. Epub. 1995/12/01.

[48] Barret D. S., Oehme, F. W., Kruckenberg, S. M. A review of organophosphorus ester induced delayed neurotoxicity. *Veterinary and Human Toxicology*. 1985;27:248-54.

[49] Vose S. C., Fujioka, K., Gulevich, A. G., Lin, A. Y., Holland, N. T., Casida, J. E. Cellular fuction of neuropathy target esterase in lysophosphatidylcholine action. *Toxicology and Applied Pharmacology*. 2008;232:376-83.

[50] Glynn P. Axonal degeneration and neuropathy target esterase. *Arh. Hig. Rada. Toksiko*. 2007;58:355-8.

[51] Greiner A. J., Richardson, R. J., Worden, R. M., Ofoli, R. Y. Influence of lysophospholipid hydrolysis by the catalytic domain of neuropathy target esterase on the fluidity of bilayer lipid membranes. *Biochemica et. Biophisica Acta*. 2010;1798:1533-9.

[52] Maroni M., Bleecker, M. L. Neuropathy target esterase in human lymphocytes and platelets. *Journal of Applied Toxicology*. 1986;6:1-7.

[53] Oliveira G. H., Moreira, V., Goes, S. P. R. Organophosphate induced delayed neuropathy in genetically dissimilar chickens: studies with tri-ortho-cresyl phosphate (T. O. C. P.) and trichlorfon. *Toxicology Letters*. 2002;136:143-50.

[54] Moretto A., Capodicasa, E., Lotti, M. Clinical expression of organophosphate-induced delayed polyneuropathy in rats. *Toxicology Letters.* 1992;63:97-102.

[55] Anthony D. C., Montine, T. J., Valentine, W. M., Graham, D. G. Toxic responses of the nervous system. . In: Klaassen C. D., editor. *Casarett and Doull Toxicology: The Basic Science of Poisons.* New York.: McGraw-Hill, 978-0-07-134721-1; 2001. p. 535–59.

[56] Wu Y. J., Li, M., Li, Y. X., Li, W., Dai, J. Y., Leng, X. F. . Verapamil abolished the enhancement of protein phosphorylation of brainstem mitochondria and synaptosomes from the hens dosed with tri-o-cresyl phosphate. *Environmental Toxicology and Pharmacology*. 2007;24: 67-71.

[57] Schlaepfer W. W., Bunge, R. P. Effects of calcium ion concentration on the degeneration of amputated axons in tissue culture. *The Journal of Cell Biology*. 1973;59:456–70.

[58] Inomata M., Hayashi, M., Nakamura, M., Imahori, K., Kawashima, S. Purification and characterization of a calcium-activated neutral protease from rabbit skeletal muscle which requires calcium ions of μM order concentration. *Journal of biochemistry*. 1983;93:291-4.

In: Enzymes and Enzyme Activity
Editor: Eva M. Lashinski

ISBN: 978-1-62417-696-8

Chapter V

Biological Effects of Calpain Inhibitors on Human Phagocyte Functions

Seiichi Kitagawa*, Takayuki Kato, Maki Kitagawa, Megumi Aomatsu and Hisakazu Fujita

Department of Physiology, Osaka City University,
Graduate School of Medicine, Osaka, Japan

Abstract

Calpain has been implicated as a regulator of actin cytoskeleton, cell migration, apoptosis, and inflammation, and calpain is a potential therapeutic target for various disorders, including neurodegenerative and inflammatory diseases.

Calpain inhibitors, including peptide aldehydes and α-mercapto-acrylic acid derivatives, are widely used as a valuable tool to elucidate the physiological and pathological roles of calpain. The biological effects of potent calpain inhibitors on human phagocytes (neutrophils and monocytes) are unique, and the inhibitors induce active migration

* Corresponding author: Seiichi Kitagawa, M.D., Department of Physiology, Osaka City University, Graduate School of Medicine, Osaka 545-8585, Japan, Tel: +81-6-6645-3715; Fax: +81-6-6645-3716, E-mail: kitagawas@med.osaka-cu.ac.jp.

(chemotaxis) of human phagocytes and delay neutrophil apoptosis with rapid activation of the signaling pathways, including Rac, mitogen-activated protein kinases, and phosphatidylinositol 3-kinase. Potent calpain inhibitors have been shown to activate human formyl peptide receptors and/or other G-protein coupled receptors in addition to inhibition of the calpain activity, which may be responsible for the characteristic effects of calpain inhibitors on human phagocyte functions.

Abbreviations

ALLM, N-acetyl-Leu-Leu-Met-CHO;
ALLN, N-acetyl-Leu-Leu-Nle-CHO;
cIAP1, cellular inhibitor of apoptosis 1;
ERK1/2, extracellular signal-regulated kinase 1/2;
fMLF, *N*-formyl-Met-Leu-Phe;
G-CSF, granulocyte colony-stimulating factor;
GM-CSF, granulocyte-macrophage CSF;
GPCR, G-protein coupled receptor;
HEK-293/hFPR, HEK-293 cells stably expressing hFPR;
HEK-293/hFPRL1, HEK-293 cells stably expressing hFPRL1;
hFPR, human formyl peptide receptor;
hFPRL1, hFPR-like 1;
IAP, inhibitor of apoptosis;
IFN-α, interferon-α;
JAK, Janus kinase;
JNK, c-Jun N-terminal kinase;
MAPK, mitogen activated protein kinase;
MCP-1, monocyte chemoattractant protein-1;
MEK1/2, MAPK/ERK kinase 1/2;
MKK3/6, MAPK kinase 3/6;
PI3K, phosphatidylinositol 3-kinase;
PKA, protein kinase A;
PTX, pertussis toxin;
STAT3, signal transducer and activator of transcription 3;
TLR, Toll-like receptor;
TNF-α, tumor necrosis factor-α;
XIAP, X-linked inhibitor of apoptosis.

Introduction

Calpain, the calcium-dependent cysteine protease, is constitutively active in resting human neutrophils and monocytes, and the calpain activity in resting neutrophils may be primarily ascribed to calpain I (μ-calpain) [1-4]. Calpain has been implicated as a regulator of cell migration and apoptosis in human neutrophils [4-6]. Calpain activation has been also shown to be involved in various disorders, including neurodegenerative diseases and myocardial dysfunction associated with myocardial infarction, and calpain is a potential therapeutic target for these disorders [7,8]. Potent calpain inhibitors are widely used for the *in vitro* and *in vivo* experiments to elucidate the role of calpain activity [3-8]. The calpain inhibitors commonly used include peptide aldehydes (such as N-acetyl-Leu-Leu-Nle-CHO [ALLN] and N-acetyl-Leu-Leu-Met-CHO [ALLM]) and α-mercapto-acrylic acid derivatives (such as PD150606 [3-(4-iodophenyl)-2-mercapto-(Z)-2-propenoic acid] and PD151746 [3-(5-fluoro-3-indolyl)-2-mercapto-(Z)-2-propenoic acid]). The inhibition of calpain activity reduces cell migration rates and invasiveness in most cell types. By contrast, potent calpain inhibitors have been shown to stimulate migration of human phagocytes (neutrophils and monocytes) [3-5]. Our recent studies show that potent calpain inhibitors not only inhibit the calpain activity but also stimulate cell functions via direct activation of human formyl peptide receptors and/or other G-protein coupled receptors (GPCRs) depending on the inhibitors used [9,10]. Here, we discuss the recent advances in the biological effects of potent calpain inhibitors on human phagocyte functions.

Neutrophil Apoptosis and Calpains

Constitutively active calpain has been implicated in regulation of neutrophil apoptosis. It has been reported that X-linked inhibitor of apoptosis (XIAP), the efficient inhibitor of caspase-3, -7, and -9 [11], can be degraded by calpain, leading to acceleration of spontaneous neutrophil apoptosis. In fact, calpain inhibitors (PD150606, ALLN, and calpeptin) can delay spontaneous and tumor necrosis factor-α (TNF-α)/cycloheximide-induced neutrophil apoptosis [12,13]. It has been also shown that granulocyte colony-stimulating factor (G-CSF) inhibits activation of calpains via apparent control of Ca^{2+} influx, leading to delayed neutrophil apoptosis [14].

Neutrophils mobilized *in vivo* by G-CSF/dexamethasone treatment exhibit a prolonged life span with up-regulation of calpastatins, the endogenous inhibitors of calpains [15]. These findings support the hypothesis that the calpain/calpastatin system may be involved in neutrophil apoptosis.

Spontaneous neutrophil apoptosis has been also shown to be closely associated with proteasome-mediated degradation of Mcl-1 and XIAP. The levels of Mcl-1 and XIAP, but not other anti-apoptotic molecules (cellular IAP1 [cIAP1], cIAP2, Bcl-2, and A1), are markedly decreased during spontaneous neutrophil apoptosis, and the reduction in the levels of Mcl-1 and XIAP as well as spontaneous neutrophil apoptosis is prevented by proteasome inhibitors (epoxomicin and lactacystin) [5,16]. Mcl-1 is stabilized by cyclic AMP and granulocyte-macrophage CSF (GM-CSF), leading to delayed neutrophil apoptosis [16,17]. Mcl-1 and XIAP are also stabilized by calpain inhibitors (PD150606 and ALLN), leading to delayed neutrophil apoptosis [5]. The stabilizing effects of calpain inhibitors on Mcl-1 and XIAP may be ascribed to cyclic AMP-independent activation of protein kinase A (PKA), since calpain inhibitors induce phosphorylation of several PKA substrates, and the effects of calpain inhibitors on phosphorylation of PKA substrates, stabilization of Mcl-1 and XIAP, and neutrophil apoptosis are suppressed by H-89, a specific inhibitor of PKA (Figure 1).

Consistent with this is the findings that the anti-apoptotic effect of calpain inhibitors and cyclic AMP on neutrophils is resistant to cycloheximide, indicating that these agents exert the anti-apoptotic effect via the protein synthesis-independent mechanism [6,18].

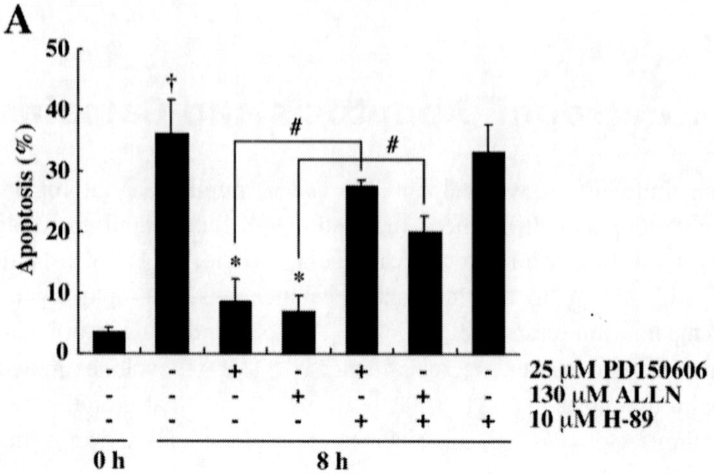

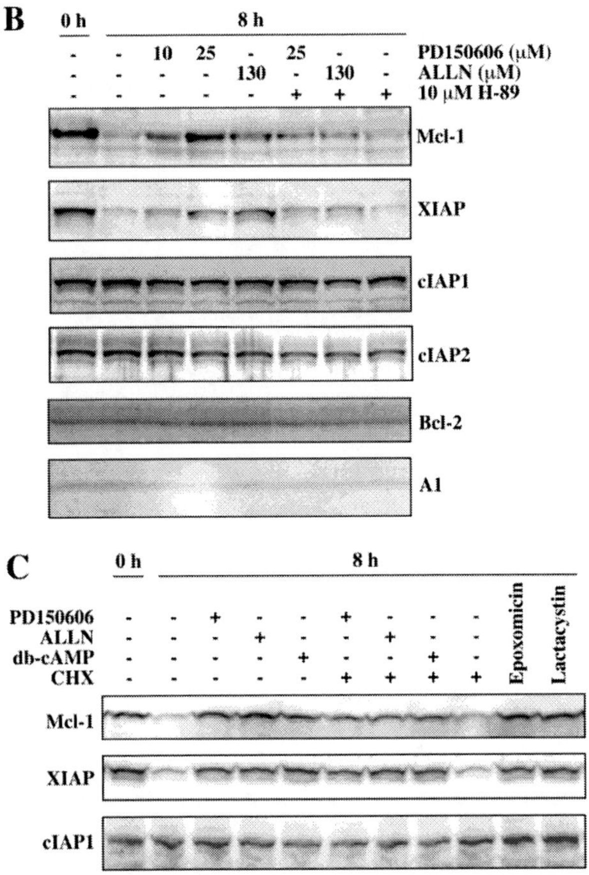

Figure 1. Calpain inhibitors delay human neutrophil apoptosis via PKA-mediated stabilization of Mcl-1 and XIAP. Neutrophils were incubated in the presence or absence of PD150606 (25 μM), ALLN (130 μM), dibutyryl cyclic AMP (db-cAMP, 1 mM), epoxomicin (1 μM), or lactacystin (5 μM) for 8 h at 37°C. When required, cells were pretreated with H-89 (10 μM) or cycloheximide (CHX, 10 μg/ml) for 30 min. (A) Apoptotic cells were analyzed by flow cytometry. Annexin V-positive, propidium iodide-negative cells were considered as apoptotic cells. †Significantly greater as compared with 0 h controls ($P < 0.01$). *Spontaneous apoptosis was significantly prevented by PD150606 or ALLN ($P < 0.01$). #PD150606- or ALLN-mediated delay of neutrophil apoptosis was significantly prevented by H-89 (PKA inhibitor) ($P < 0.05$). (B and C) The immunoblotting was performed using antibodies against each protein. The levels of Mcl-1 and XIAP were markedly decreased during spontaneous neutrophil apoptosis, and the reduction in the levels of Mcl-1 and XIAP was prevented by calpain inhibitors (PD150606 and ALLN), proteasome inhibitors (epoxomicin and lactacystin), or dibutyryl cyclic AMP. The stabilizing effect of PD150606 and ALLN on Mcl-1 and XIAP was suppressed by H-89. (Reproduced from [6]).

Mitogen activated protein kinases (MAPKs), including extracellular signal-regulated kinase 1/2 (ERK1/2), p38, and c-Jun N-terminal kinase (JNK), and phosphatidylinositol 3-kinase (PI3K)/Akt are activated in human neutrophils upon exposure to calpain inhibitors. Thus, calpain inhibitors might delay neutrophil apoptosis not only by inhibiting the calpain activity but also by activating pro-survival molecules such as MEK/ERK and PI3K/Akt, both of which are known to be involved in delayed neutrophil apoptosis under certain situations. However, the latter possibility appears to be unlikely, as we observed that MEK and PI3K inhibitors did not affect calpain inhibition-mediated delay of spontaneous neutrophil apoptosis [4,5].

Accumulated evidences suggest that degradation of Mcl-1 and XIAP by calpain and/or proteasome may play an important role in spontaneous neutrophil apoptosis. However, it should be noted that potent calpain inhibitors commonly used could inhibit not only the calpain activity but also the proteasome activity, and activate human formyl peptide receptors [9,19], indicating that careful interpretation is required when these inhibitors are used.

Spontaneous neutrophil apoptosis is also modulated by inflammatory cytokines, such as G-CSF, GM-CSF, interferon-α (IFN-α), IFN-γ, and TNF-α [20]. G-CSF, GM-CSF, IFN-α, and IFN-γ delay neutrophil apoptosis, whereas TNF-α accelerates neutrophil apoptosis especially in the presence of cycloheximide [18, 21, 22].

Modulation of neutrophil apoptosis by inflammatory cytokines may be closely associated with the outcome of inflammation. Neutrophil apoptosis may be regulated by up- or down-regulation of molecules associated with apoptosis, which may include the Bcl-2 family proteins (Bcl-X_L, Mcl-1, A1, and Bax) and the inhibitor of apoptosis (IAP) proteins (cIAP2, XIAP, and survivin). For example, it has been reported that G-CSF up-regulates the expression of A1, cIAP2, and survivin, and down-regulates the expression of Bax [15,23].

GM-CSF up-regulates the expression of Mcl-1 and survivin, stabilizes Mcl-1, and down-regulates the expression of Bax-α [23-26], whereas TNF-α down-regulates the expression of Bcl-X_L. IFN-α and IFN-γ up-regulate the expression of cIAP2 [21]. In regard to the signaling pathways, tyrosine kinase Lyn, ERK, PI3K, and Janus kinase (JAK)/signal transducer and activator of transcription 3 (STAT3) have been proposed to be involved in the G-CSF-, GM-CSF-, IFN-α-, or IFN-γ-mediated anti-apoptotic effect on human neutrophils [17,21,22,27].

Neutrophil Migration and Calpain Inhibitors

It is generally believed that calpain promotes the turnover of adhesion complexes by cleaving focal adhesion proteins such as focal adhesion kinase, talin, and paxillin. Calpain-deficient (Capn4$^{-/-}$; a small subunit-deficient) fibroblasts exhibit reduced migration, extended numerous thin projections, and delayed retraction of membrane protrusions and filopodia [28]. In contrast to all other cell types examined, inhibition of calpain I (μ-calpain) activity promotes migration in human neutrophils [4,5]. Lokuta et al. [4] have demonstrated that active calpain I is expressed predominantly and constitutively in resting human neutrophils, and the specific inhibitor for calpain I but not calpain II (m-calpain) induces neutrophil polarization and migration. By contrast, chemoattractants induce asymmetric recruitment of calpain II (m-calpain), but not calpain I, to the leading edge of polarized human neutrophils, and calpain II may play a role in regulating pseudopodia formation [29].

MAPKs, including ERK1/2, p38 and JNK, MAPK/ERK kinase 1/2 (MEK1/2), MAPK kinase 3/6 (MKK3/6), PI3K/Akt, c-Raf, and p21-activated kinase (PAK, an effector molecule of Rac) are rapidly (within 30 s) activated in human neutrophils upon exposure to calpain inhibitors (PD150606 and ALLN), but not PD145305 (2-mercapto-3-phenylpropanonic acid; inactive analog of PD150606). Following activation of these pathways, neutrophils display active migration, which is sustained for more than 45 min. The micropipette method reveals that calpain inhibition-mediated neutrophil migration is chemotaxis, but not random migration (Figure 2). The studies with pharmacological inhibitors indicate that calpain inhibition-mediated neutrophil migration may be mediated by activation of MEK/ERK, p38, JNK, PI3K/Akt, and Rac [5].

NSC23766 (Rac inhibitor) and pertussis toxin (PTX) suppress calpain inhibitor-induced phosphorylation of distinct signaling molecules (PAK, c-Raf, MEK1/2, ERK, MKK3/6, p38, JNK, and Akt) as well as cell migration, suggesting that the PTX-sensitive G protein and Rac axis may be a possible key target of calpain inhibitors. Differentiated neutrophil-like HL-60 cells, but not undifferentiated cells, display cell migration and activation of MAPKs and PI3K/Akt on calpain inhibition, indicating that mature neutrophils specifically respond to calpain inhibitors [5]. Possible involvement of MAPKs and PI3K in neutrophil migration has been shown in several experimental systems, although the roles of these molecules may be dependent on the agonists used [30-35].

MEK/ERK, p38, and PI3K/Akt are involved in actin reorganization in human neutrophils stimulated with G-CSF, GM-CSF, TNF-α, or Toll-like receptor (TLR) agonists [30-32,34].

MEK/ERK is involved in G-CSF-induced neutrophil random migration, and is associated with G-CSF-induced redistribution of F-actin and phosphorylated myosin light chain [30,32]. Coordinated action of MEK/ERK and p38 is involved in neutrophil random migration induced by TLR2 and TLR4 agonists [31].

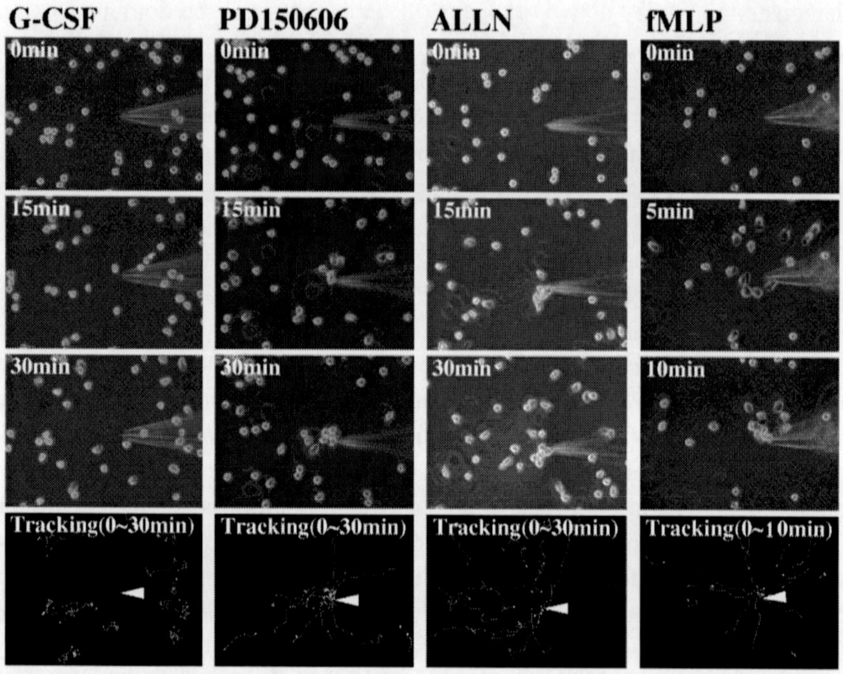

Figure 2. Calpain inhibition induces human neutrophil chemotaxis. G-CSF (2.5 μg/ml), PD150606 (500 μM), ALLN (6.5 mM), or fMLP (= fMLF, 1 μM) was loaded into a sterile injection capillary, Femtotip II. A gradient was formed by slow release of the stimuli from the tip into the medium by using an Eppendorf microinjector. The images at the indicated time-points after the release of G-CSF, PD150606, ALLN, or fMLP are shown. The tracks that neutrophils moved along during the indicated periods are shown at the bottom. An arrowhead indicates the micropipette tip. Note that PD150606, ALLN, and fMLP induce neutrophil chemotaxis, whereas G-CSF induces neutrophil random migration. (Reproduced from [5]).

Activation of p38 and MAPK activated protein kinase-2, a downstream kinase of p38, plays a role in *N*-formyl-Met-Leu-Phe (fMLF)-induced chemotaxis [35].

Monocyte Migration and Calpain Inhibitors

Calpain is also constitutively active in resting human monocytes, but not lymphocytes, and human monocytes, but not lymphocytes, exhibit active migration in response to potent calpain inhibitors [3]. As observed in human neutrophils, MAPKs (ERK, p38, and JNK), PI3K/Akt, and PAK are rapidly (within 1 min) activated in human monocytes, but not lymphocytes, upon exposure to calpain inhibitors (PD150606 and ALLN), but not PD145305. Following activation of these signalling pathways, monocytes display active migration within 5 min after exposure to calpain inhibitors, and active migration is sustained for more than 45 min. The micropipette method reveals that calpain inhibition-mediated monocyte migration is chemotaxis, but not random migration.

The studies with pharmacological inhibitors suggest that calpain inhibition-mediated monocyte migration is mediated by activation of ERK, p38, JNK, PI3K/Akt, and Rac. NSC23766 and PTX suppress calpain inhibitor-induced phosphorylation of distinct signaling molecules (PAK, ERK, p38, JNK, and Akt) as well as cell migration, suggesting that the PTX-sensitive G protein and Rac axis may be a possible key target of calpain inhibitors [3]. Thus, the similar pathways are activated in human neutrophils and monocytes when cells are exposed to potent calpain inhibitors.

Possible involvement of MAPKs and PI3K in monocyte migration has been shown in several experimental systems, although the roles of these molecules may be dependent on the agonists used and some controversial results are reported [36-40]. MEK/ERK and PI3K are shown to be involved in monocyte chemotaxis induced by monocyte chemoattractant protein-1 (MCP-1) [36]. JNK and p38, but not ERK, are involved in MCP-1-mediated trans-endothelial migration of MonoMac6 cells, a monocytic cell line [37]. ERK and p38 are involved in monocyte migration induced by angiotensin II, whereas fMLF-induced monocyte transmigration is independent of activation of ERK and p38 [38,39]. It has been also reported that neither ERK nor PI3K are involved in monocyte chemotaxis induced by MCP-1 or fMLF [40]. The controversial results may be partly ascribed to the different experimental methods or the different cell types (primary monocytes or a monocytic cell line) used. It should be also noted that MAPKs can be differentially involved in cell migration according to the agonists used [5,30,31].

Calpain Inhibitors Activate Human Formyl Peptide Receptors

Calpain inhibitors induce human neutrophil and monocyte chemotaxis, which is sensitive to PTX [3,5]. PTX catalyzes ADP-ribosylation of specific G-protein α subunit (Giα), leading to prevention of the interaction between G-protein and certain GPCRs [41]. The responses of neutrophils and monocytes to calpain inhibitors apparently mimic those to chemoattractants such as fMLF and IL-8, raising the possibility that certain PTX-sensitive GPCRs such as formyl peptide receptors may be directly or indirectly involved in cell activation by calpain inhibitors. The formyl peptide receptor family plays an important role in host defense against pathogens, but also in sensing internal molecules such as mitochondrial proteins that may be associated with cellular dysfunction. In humans, there are three types of receptors; human formyl peptide receptor (hFPR), hFPR-like 1 (hFPRL1), and hFPRL2. hFPR and hFPRL1 are expressed on various types of cells, including neutrophils, monocytes/macrophages, immature dendritic cells, endothelial cells, platelets, hepatocytes, astrocytes, microglial cells, fibroblasts, and vascular smooth muscle cells, whereas hFPRL2 is expressed on monocytes/macrophages and dendritic cells [42].

Our recent studies have demonstrated that various calpain inhibitors (PD150606, PD151746, ALLN, ALLM, and calpeptin) and γ-secretase inhibitor I induce PTX-sensitive increase in cytoplasmic free Ca^{2+} ($[Ca^{2+}]_i$) in human neutrophils and neutrophil migration. In addition, HEK-293 cells stably expressing hFPR (HEK-293/hFPR) or hFPRL1 (HEK-293/hFPRL1) display stimulus-specific increase in $[Ca^{2+}]_i$ in response to calpain inhibitors (PD150606, PD151746, ALLN, ALLM, MG-132, and calpeptin), γ-secretase inhibitor I, and fMLF [9]. Interestingly, the peptide inhibitors of calpain (ALLN and ALLM) may activate both hFPR and hFPRL1, as evidenced by the findings that ALLN and ALLM induce an increase in $[Ca^{2+}]_i$ and ERK phosphorylation in HEK-293/hFPR and HEK-293/hFPRL1, but not parent HEK-293 cells. On the other hand, α-mercapto-acrylic acid derivatives

(PD150606 and PD151746) may activate hFPRL1, but not hFPR, as evidenced by the findings that PD150606 and PD151746 induce an increase in $[Ca^{2+}]_i$ and ERK phosphorylation in HEK-293/hFPRL1, but not HEK-293/hFPR and parent HEK-293 cells. These findings may provide the molecular mechanisms for explanation of the observations that certain calpain inhibitors such as ALLN and PD150606 can induce human neutrophil or monocyte migration [3-5], and may also explain why fMLF-induced neutrophil chemotaxis is disturbed by pretreatment of cells with ALLN or PD150606 [4].

The peptide aldehydes such as ALLN and ALLM exhibit competitive inhibition with respect to substrate and interact with the active sites of calpain, whereas α-mercapto-acrylic acid derivatives exhibit uncompetitive inhibition and interact with the calcium-binding domain of the large and small subunits of calpain, but not the active sites [43,44]. These different characteristics of the peptide aldehydes and α-mercapto-acrylic acid derivatives may be associated with the different activation of hFPR and/or hFPRL1 by these inhibitors. In this regard, it has been reported that not only the *N*-formylated form but also the acetylated form of the peptide can activate hFPR [45], and a substituted quinazolinone (Quin-C1) and a lipid product (lipoxin A4) can activate hFPRL1 [46,47].

Both calpeptin (Z-Leu-Nle-CHO) and γ-secretase inhibitor I (Z-Leu-Leu-Nle-CHO) induce an increase in $[Ca^{2+}]_i$ in parent HEK-293 cells and neutrophils, and induce neutrophil migration. Furthermore, calpeptin- or γ-secretase inhibitor I-induced increase in $[Ca^{2+}]_i$ in parent HEK-293 cells and neutrophils is sensitive to PTX.

These findings suggest that calpeptin and γ-secretase inhibitor I could stimulate cells via activation of other PTX-sensitive GPCRs than hFPR and hFPRL1, and these receptors are shared by HEK-293 cells and human neutrophils. MG-132 (Z-Leu-Leu-Leu-CHO) also induces an increase in $[Ca^{2+}]_i$ in parent HEK-293 cells and neutrophils. However, MG-132, unlike calpeptin and γ-secretase inhibitor I, fails to induce neutrophil migration, and MG-132-induced increase in $[Ca^{2+}]_i$ in parent HEK-293 cells and neutrophils is resistant to PTX. These findings suggest that MG-132 stimulates cells via activation of other receptors than hFPR and hFPRL1, and these receptors are resistant to PTX and shared by HEK-293 cells and human neutrophils. Possible candidates may include Gq-coupled GPCRs, which are involved in an increase in $[Ca^{2+}]_i$, but not cell migration [41,48]. In this regard, it is of interest that MG-132 has been shown to stimulate various cell functions via activation of the distinct signaling pathways.

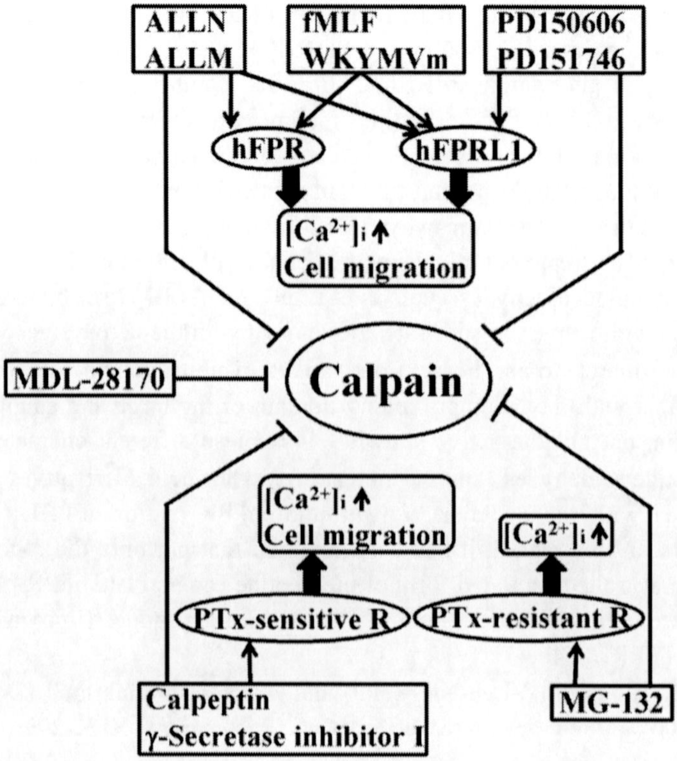

Figure 3. Possible targets of calpain inhibitors. Potent calpain inhibitors activate human formyl peptide receptors (hFPR and hFPRL1) and/or other GPCRs in addition to inhibition of the calpain activity, which may be responsible for the characteristic effects of calpain inhibitors on human phagocyte functions. PTx-sensitive R, pertussis toxin-sensitive receptor.

For example, MG-132 induces IL-6 expression via activation of MEK/ERK in human umbilical vein endothelial cells [49], and IL-8 expression via activation of ERK and JNK in human epithelial A549 cells [50], and prolongs mouse osteoclast survival via activation of PI3K/Akt [51]. These effects of MG-132 might be mediated by activation of the same receptors as those shared by HEK-293 cells and human neutrophils, but not by inhibition of the calpain activity. MDL-28170 (Z-Val-Phe-CHO) induces neither an increase in $[Ca^{2+}]_i$ in neutrophils and HEK-293 cells nor neutrophil migration. Nevertheless, MDL-28170 induces phosphorylation of ERK, p38, and Akt in human neutrophils, suggesting that MDL-28170-mediated activation of these molecules might be associated with inhibition of the calpain activity. Ionomycin-induced cleavage of talin (a substrate of calpain) in neutrophils is inhibited by all inhibitors described above, indicating that all these inhibitors can inhibit the calpain activity. The $[Ca^{2+}]_i$ responses in human neutrophils and HEK-293 cells stimulated by various inhibitors or ligands are summarized in Table 1, and possible targets of potent calpain inhibitors are shown in Figure 3 [9]. It should be noted that potent calpain inhibitors not only inhibit the calpain activity but also activate hFPR, hFPRL1, and/or other GPCRs, leading to activation of phagocyte functions and modulation of neutrophil survival.

Table 1. $[Ca^{2+}]_i$ responses in human neutrophils and HEK-293 cells stimulated by various inhibitors or ligands

Inhibitors/ Ligands	Chemical structures	Increase in $[Ca^{2+}]_i$			
		Neutrophils	HEK-293/ hFPR	HEK-293/ hFPRL1	HEK-293/ empty
PD150606*		+	-	+	-
PD151746*		+	-	+	-
PD145305*		-	-	-	-
ALLN	N-Ac-Leu-Leu-Nle-CHO	+	+	+	-
ALLM	N-Ac-Leu-Leu-Met-CHO	+	+	+	-
MDL-28170	Z-Val-Phe-CHO	-	-	-	-
MG-132	Z-Leu-Leu-Leu-CHO	+†	+	+	+†
Calpeptin	Z-Leu-Nle-CHO	+	+	+	+
γ-Secretase inhibitor I	Z-Leu-Leu-Nle-CHO	+	+	+	+
fMLF	N-formyl-Met-Leu-Phe	+	+	+	-
WKYMVm	Trp-Lys-Tyr-Met-Val-D-Met	+	+	+	-

Synonyms: ALLN, calpain inhibitor I and MG-101; ALLM, calpain inhibitor II; MDL-28170, calpain inhibitor III; MG-132, calpain inhibitor IV; WKYMVm, W peptide.

N-Ac, N-acetyl; Z, N-benzyloxycarbonyl; CHO, aldehyde.

*PD150606 ($C_9H_7IO_2S$), PD151746 ($C_{11}H_8FNO_2$), and PD145305 ($C_9H_{10}O_2S$) are α-mercapto-acrylic acid derivatives.

†MG-132-induced increase in $[Ca^{2+}]_i$ in neutrophils and parent HEK-293 cells is resistant to PTX.

Active migration of neutrophils and monocytes exposed to potent calpain inhibitors suggest that the calpain activity may be dispensable for neutrophil and monocyte migration. In fact, fMLF-induced neutrophil migration is unaffected by MDL-28170, which can inhibit the calpain activity, but cannot induce an increase in $[Ca^{2+}]_i$, in human neutrophils [9]. When peptide aldehydes and α-mercapto-acrylic acid derivatives are employed *in vivo* or *in vitro*, it would be expected that the functions of hFPR- and/or hFPRL1-expressing cells would be affected. hFPR and hFPRL1 are expressed on various types of cells [42], and have been also shown to be expressed on malignant cells such as glioma and breast tumor cell lines, and to be involved in the invasiveness or proliferation of these cells [52,53]. These findings indicate that possible expression of hFPR and/or hFPRL1 on primary tumor cells should be examined in case of clinical application of, at least, certain types of calpain inhibitors.

Homology Modeling of Human Formyl Peptide Receptors and Ligand Docking Simulation

Recent studies indicate that potent calpain inhibitors not only inhibit the calpain activity but also stimulate cell functions via direct activation of hFPR, hFPRL1, and/or other receptors depending on the inhibitors used [9]. Specifically, ALLN and ALLM, like fMLF, may induce phagocyte (neutrophil and monocyte) migration via activation of hFPR and hFPRL1, whereas PD150606 and PD151746 may induce phagocyte migration via activation of hFPRL1, but not hFPR [9]. Using the homology modeling of the receptors and the ligand docking simulation, we found that these calpain inhibitors could bind to the putative fMLF binding site on hFPR and/or hFPRL1 [10].

We initially constructed the three-dimensional structure models of hFPR and hFPRL1 by the homology modeling using the Phyre server, and identified the possible fMLF binding site on these receptors. The ligand docking simulation reveals that ALLN could bind to the putative fMLF binding site on hFPR and hFPRL1 with high probability, and PD150606 could bind to the putative fMLF binding site on hFPRL1, but not hFPR, with high probability (Figure 4) [10]. These findings are consistent with the ability of these compounds to induce an increase in $[Ca^{2+}]_i$ and activation of ERK in HEK-293/hFPR and/or HEK-293/hFPRL1 [9].

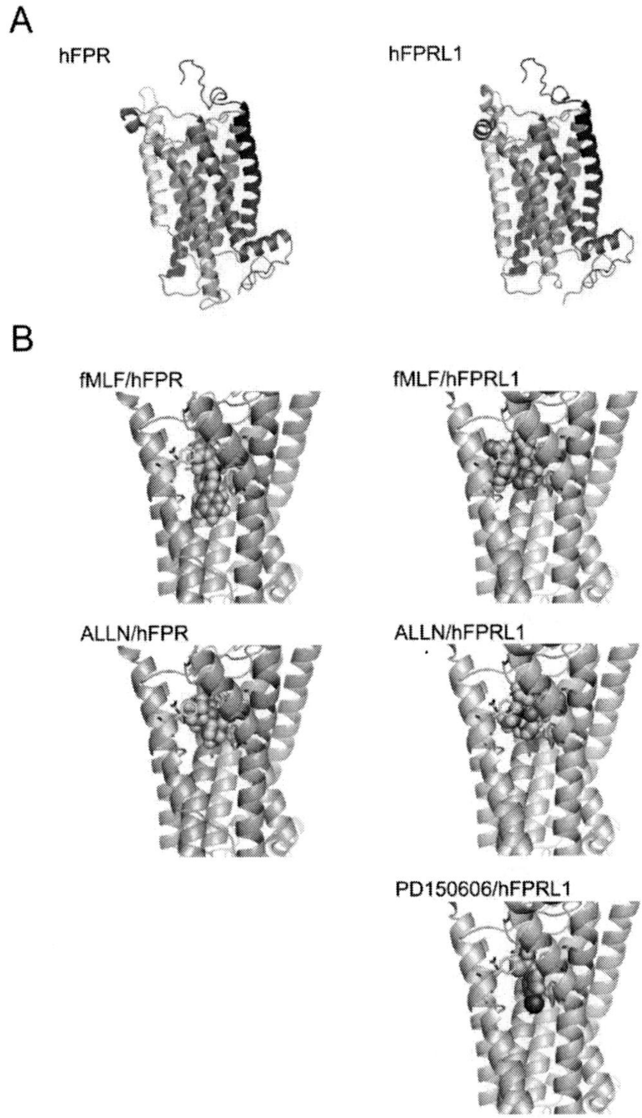

Figure 4. Homology modeling of hFPR and hFPRL1, and ligand docking simulation. (A) The predicted models of hFPR and hFPRL1. The trans-membrane (TM) helices are colored as follows: TM1 (blue), TM2 (cyan), TM3 (green), TM4 (light green), TM5 (yellow), TM6 (orange), and TM7 (red). (B) The binding of fMLF, ALLN, and PD150606 to hFPR or hFPRL1. The ligand docking model with the lowest interaction energy is shown for each binding. The helices are shown in gray with the side chains of the amino acids in color. Ligands are shown by the space filling mode. (Reproduced from [10]).

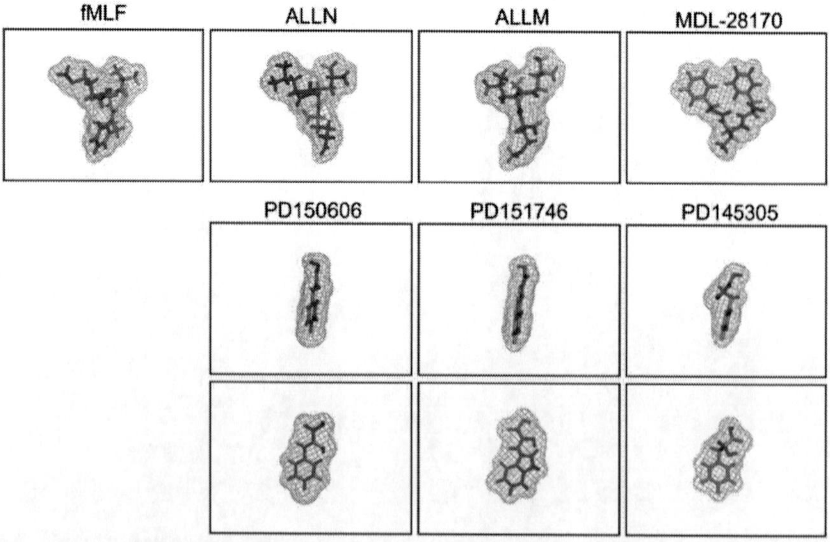

Figure 5. Three-dimensional structure of fMLF, ALLN, ALLM, MDL-28170, PD150606, PD151746, and PD145305. The different views are shown for PD150606, PD151746, and PD145305. Note that the three-dimensional structure of ALLN and ALLM, but not MDL-28170, is similar to that of fMLF, whereas the three-dimensional structures of the peptide aldehydes and α-mercapto-acrylic acid derivatives are apparently different from each other. (Reproduced from [10]).

The ligand docking simulation also reveals that MDL-28170 and PD145305 are unable to bind to the putative fMLF binding site on hFPR and hFPRL1. The three-dimensional structure of MDL-28170 is apparently different from that of fMLF, ALLN, or ALLM, and the three-dimensional structure of PD145305 is apparently different from that of PD150606 or PD151746 (Figure 5). These findings are consistent with the observations that neither MDL-28170 nor PD145305 induce an increase in $[Ca^{2+}]_i$ in human neutrophils, HEK-293/hFPR, and HEK-293/hFPRL1 [9,10].

The three-dimensional structure of ALLN and ALLM is similar to that of fMLF, a finding consistent with the notion that ALLN and ALLM, like fMLF, may activate hFPR and hFPRL1 by binding to the putative fMLF-binding sites on the receptors (Figure 5).

However, the three-dimensional structures of the peptide aldehydes and α-mercapto-acrylic acid derivatives are apparently different from each other, and these calpain inhibitors are known to inhibit the calpain activity through the different mechanisms [43,44]. Nevertheless, both types of inhibitors (ALLN and PD150606) bind to the putative fMLF binding site on hFPRL1 with high probability.

In addition, both types of inhibitors inhibit the calpain activity in ionomycin-stimulated neutrophils at the same concentration range as that required to induce an increase in $[Ca^{2+}]_i$. By contrast, PD150606, unlike ALLN, could not bind to the putative fMLF binding site on hFPR. These findings indicate that these calpain inhibitors can inhibit the calpain activity in intact cells, and, at the same time, activate hFPR and/or hFPRL1 via direct binding to the putative fMLF binding site, which is apparently independent of the ability of these inhibitors to inhibit the calpain activity [9].

The lack of α-mercapto-acrylic acid derivatives (PD150606 and PD151746) to bind to hFPR may be ascribed to the structural difference between hFPR and hFPRL1; the amino acid sequence of hFPRL1 being 69% identical to that of hFPR [54]. It has been reported that PD150606 binds to the hydrophobic pocket in the domain VI of the small subunit of calpain, which is formed by three helices; E1, F1, and F2E3 [55].

The ligand docking simulation reveals that PD150606 could bind to the hydrophobic pocket in hFPRL1 formed by trans-membrane helix 3 (TMH3), TMH5, TMH6, and TMH7. TMH3 of hFPR contains Phe102 (a non-polar amino acid), whereas TMH3 of hFPRL1 contains His102 (a basic amino acid). Interestingly, PD150606 and PD151746 contain iodine and fluorine, respectively. These findings taken together suggest that high electronegativity of iodine and fluorine might partly explain why PD150606 and PD151746 could bind to hFPRL1, but not hFPR [9].

The studies with HEK-293/hFPR and HEK-293/hFPRL1 show that the concentrations of calpain inhibitors required to induce an increase in $[Ca^{2+}]_i$ (> 0.1 μM for PD150606 and > 1 μM for ALLN, respectively) is much higher (> 100 folds) than those of fMLF and Trp-Lys-Tyr-Met-Val-D-Met (WKYMVm, W peptide), a ligand that preferentially stimulates hFPRL1. HEK-293 cells expressing hFPR or hFPRL1 with the mutated fMLF binding site never exhibit the $[Ca^{2+}]_i$ response to calpain inhibitors. When the optimal concentrations of each stimulus are used, pretreatment of cells with fMLF or WKYMVm abolishes an increase in $[Ca^{2+}]_i$ induced by calpain inhibitors as well as the same stimulus, whereas pretreatment of cells with calpain inhibitors signify-cantly suppresses, but never abolish, the $[Ca^{2+}]_i$ response induced by fMLF or WKYMVm, suggesting that the binding affinity of the inhibitors (PD150606 and ALLN) to the putative fMLF binding site may be lower than that of fMLF or WKYMVm [10]. It should be noted that the concentrations of the calpain inhibitors used in various studies are apparently sufficient to activate hFPR, hFPRL1, and/or other GPCRs. For example, calpeptin (100 μM) and PD150606 (50 μM) are used as calpain inhibitors to prevent XIAP degradation and to delay human neutrophil apoptosis [13]. ALLN (100 μM) and MG-132 (50 μM) are used as proteasome inhibitors to prevent Mcl-1 degradation and to delay human neutrophil apoptosis [17].

Conclusion

ALLN and ALLM are originally reported to be potent inhibitors of calpains and cathepsins [43], and thereafter these peptide aldehydes, including ALLN, ALLM, and MG-132, have been shown to be capable of inhibiting proteasome [19]. These peptide aldehydes are now widely used as a valuable tool to elucidate the roles of calpain and proteasome. Recent studies have demonstrated that these peptide aldehydes not only exert the inhibitory effects on calpain and proteasome, but also can stimulate cell functions via activation of hFPR, hFPRL1, and/or other GPCRs depending on the cell types used [9,10]. Therefore, the *in vitro* and *in vivo* experiments with potent calpain inhibitors should be carefully interpreted. In addition, further studies are required to elucidate the physiological roles of constitutively active calpain in resting human neutrophils and monocytes. In any case, the recent advances in the biological effects of potent calpain inhibitors discussed here may be helpful for interpretation of the experimental results with calpain inhibitors and development of useful calpain inhibitors for the therapeutics.

Disclosure

The authors have no financial conflicts of interest.

References

[1] Franco, S. J., Huttenlocher, A. (2005) Regulating cell migration: calpains make the cut. *J. Cell Sci.* 118: 3829–3838.

[2] Wells, A., Huttenlocher, A., Lauffenburger, D. A. (2005) Calpain proteases in cell adhesion and motility. *Int. Rev. Cytol.* 245: 1-16.

[3] Noma, H., Kato, T., Fujita, H., Kitagawa, M., Yamano, T., Kitagawa, S. (2009) Calpain inhibition induces activation of the distinct signalling pathways and cell migration in human monocytes. *Immunology* 128: e487-e496.

[4] Lokuta, M. A., Nuzzi, P. A., Huttenlocher, A. (2003) Calpain regulates neutrophil chemotaxis. *Proc. Natl. Acad. Sci. USA* 100: 4006-4011.

[5] Katsube, M., Kato, T., Kitagawa, M., Noma, H., Fujita, H., Kitagawa, S. (2008) Calpain-mediated regulation of the distinct signaling pathways and cell migration in human neutrophils. *J. Leukoc. Biol.* 84: 255-263.

[6] Ozaki, Y., Kato, T., Kitagawa, M., Fujita, H., Kitagawa, S. (2008) Calpain inhibition delays neutrophil apoptosis via cyclic AMP-independent activation of protein kinase A and protein kinase A-mediated stabilization of Mcl-1 and X-linked inhibitor of apoptosis (XIAP). *Arch. Biochem. Biophys.* 477: 227-231.

[7] Camins, A., Verdaguer, E., Folch, J., Pallas, M. (2006) Involvement of calpain activation in neurodegenerative processes. *CNS Drug Rev.* 12: 135-148.

[8] Mani, S. K., Balasubramanian, S., Zavadzkas, J. A., Jeffords, L. B., Rivers, W. T., Zile, M. R., Mukherjee, R., Spinale, F. G., Kuppuswamy, D. (2009) Calpain inhibition preserves myocardial structure and function following myocardial infarction. *Am. J. Physiol. Heart Circ. Physiol.* 297: H1744-H1751.

[9] Fujita, H., Kato, T., Watanabe, N., Takahashi, T., Kitagawa, S. (2011) Calpain inhibitors stimulate phagocyte functions via activation of human formyl peptide receptors. *Arch. Biochem. Biophys*. 513: 51-60.

[10] Fujita, H., Kato, T., Watanabe, N., Takahashi, T., Kitagawa, S. (2011) Stimulation of human formyl peptide receptors by calpain inhibitors: Homology modeling of receptors and ligand docking simulation. *Arch. Biochem. Biophys*. 516: 121-127.

[11] Deveraux, Q. L., Roy, N., Stennicke, H. R., Van Arsdale, T., Zhou, Q., Srinivasula, S. M., Alnemri, E. S., Salvesen, G. S., Reed, J. C. (1998) IAPs block apoptotic events induced by caspase-8 and cytochrome c by direct inhibition of distinct caspases. *EMBO J.* 17: 2215-2223.

[12] Knepper-Nicolai, B., Savill, J., Brown, S. B. (1998) Constitutive apoptosis in human neutrophils requires synergy between calpains and the proteasome downstream of caspases. *J. Biol. Chem.* 273: 30530-30536.

[13] Kobayashi, S., Yamashita, K., Takeoka, T., Ohtsuki, T., Suzuki, Y., Takahashi, R., Yamamoto, K., Kaufmann, S. H., Uchiyama, T., Sasada, M., Takahashi, A. (2002) Calpain-mediated X-linked inhibitor of apoptosis degradation in neutrophil apoptosis and its impairment in chronic neutrophilic leukemia. *J. Biol. Chem.* 277: 33968-33977.

[14] van Raam, B. J., Drewniak, A., Groenewold, V., van den Berg, T. K., Kuijpers, T. W. (2008) Granulocyte colony-stimulating factor delays neutrophil apoptosis by inhibition of calpains upstream of caspase-3. *Blood* 112: 2046-2054.

[15] Drewniak, A, van Raam, B. J., Geissler, J., Tool, A. T., Mook, O. R., van den Berg, T. K., Baas, F., Kuijpers, T. W. (2009) Changes in gene expression of granulocytes during in vivo granulocyte colony-stimulating factor/dexamethasone mobilization for transfusion purposes. *Blood* 113: 5979-5998.

[16] Kato, T., Kutsuna, H., Oshitani, N., Kitagawa, S. (2006) Cyclic AMP delays neutrophil apoptosis via stabilization of Mcl-1. *FEBS Lett.* 580: 4582-4586.

[17] Derouet, M., Thomas, L., Cross, A., Moots, R. J., Edwards, S. W. (2004) Granulocyte macrophage colony-stimulating factor signaling and proteasome inhibition delay neutrophil apoptosis by increasing the stability of Mcl-1. *J. Biol. Chem.* 279: 26915-26921.

[18] Sakamoto, C., Suzuki, K., Hato, F., Akahori, M., Hasegawa, T., Hino, M., Kitagawa, S. (2003) Anti-apoptotic effect of granulocyte colony-stimulating factor, granulocyte-macrophage colony-stimulating factor and cyclic AMP on human neutrophils: protein synthesis-dependent and protein synthesis-independent mechanisms and role of Janus kinase-STAT pathway. *Int. J. Hematol.* 77: 60-70.

[19] Rock, K. L., Gramm, C., Rothstein, L., Clark, K., Stein, R., Dick, L., Hwang, D., Goldberg, A. L. (1994) Inhibitors of the proteasome block the degradation of most cell proteins and the generation of peptides presented on MHC class I molecules. *Cell* 78: 761-771.

[20] Kato, T., Kitagawa, S. (2006) Regulation of neutrophil functions by proinflammatory cytokines. *Int. J. Hematol.* 84: 205-209. Review.

[21] Sakamoto, E., Hato, F., Kato, T., Sakamoto, C., Akahori, M., Hino, M., Kitagawa, S. (2005) Type I and type II interferons delay human neutrophil apoptosis via activation of STAT3 and up-regulation of cellular inhibitor of apoptosis 2. *J. Leukoc. Biol.* 78: 301-309.

[22] Hasegawa, T., Suzuki, K., Sakamoto, C., Ohta, K., Nishiki, S., Hino, M., Tatsumi, N., Kitagawa, S. (2003) Expression of the inhibitor of apoptosis (IAP) family members in human neutrophils: up-regulation of cIAP2 by granulocyte colony-stimulating factor and overexpression of cIAP2 in chronic neutrophilic leukemia. *Blood* 101: 1164-1171.

[23] Altznauer, F., Martinelli, S., Yousefi, S., Thürig, C., Schmid, I., Conway, E. M., Schöni, M. H., Vogt, P., Mueller, C., Fey, M. F., Zangemeister-Wittke, U., Simon, H. U. (2004) Inflammation-associated cell cycle-independent block of apoptosis by survivin in terminally differentiated neutrophils. *J. Exp. Med.* 199: 1343-1354.

[24] Moulding, D. A., Quayle, J. A., Hart, A., Edwards, S. W. (1998) Mcl-1 expression in human neutrophils: regulation by cytokines and correlation with cell survival. *Blood* 92: 2495-2502.

[25] Moulding, D. A., Akgul, C., Derouet, M., White, M. R. H., Edwards, S. W. (2001) BCL-2 family expression in human neutrophils during delayed and accelerated apoptosis. *J. Leukoc. Biol.* 70: 783-792.

[26] Epling-Burnette, P. K., Zhong, B., Bai, F., Jiang, K., Bailey, R. D., Garcia, R., Jove, R., Djeu, J. Y., Loughran, T. P. Jr., Wei, S. (2001) Cooperative regulation of Mcl-1 by Janus kinase/stat and phosphatidylinositol 3-kinase contribute to granulocyte-macrophage colony-stimulating factor-delayed apoptosis in human neutrophils. *J. Immunol.* 166: 7486-7495.

[27] Klein, J. B., Rane, M. J., Scherzer, J. A., Coxon, P. Y., Kettritz, R., Mathiesen, J. M., Buridi, A., McLeish, K. R. (2000) Granulocyte-macrophage colony-stimulating factor delays neutrophil constitutive apoptosis through phosphoinositide 3-kinase and extracellular signal-regulated kinase pathways. *J. Immunol.* 164: 4286-4291.

[28] Dourdin, N., Bhatt, A. K., Dutt, P., Greer, P. A., Arthur, J. S., Elce, J. S., Huttenlocher, A. (2001) Reduced cell migration and disruption of the actin cytoskeleton in calpain-deficient embryonic fibroblasts. *J. Biol. Chem.* 276: 48382-48388.

[29] Nuzzi, P. A., Senetar, M. A., Huttenlocher, A. (2007) Asymmetric localization of calpain 2 during neutrophil chemotaxis. *Mol. Biol. Cell* 18: 795-805.

[30] Nakamae-Akahori, M., Kato, T., Masuda, S., Sakamoto, E., Kutsuna, H., Hato, F., Nishizawa, Y., Hino, M., Kitagawa, S. (2006) Enhanced neutrophil motility by granulocyte colony-stimulating factor: the role of extracellular signal-regulated kinase and phosphatidylinositol 3-kinase. *Immunology* 119: 393-403.

[31] Aomatsu, K., Kato, T., Fujita, H., Hato, F., Oshitani, N., Kamata, N., Tamura, T., Arakawa, T., Kitagawa, S. (2008) Toll-like receptor agonists stimulate human neutrophil migration via activation of mitogen-activated protein kinases. *Immunology* 123: 171-180.

[32] Kutsuna, H., Suzuki, K., Kamata, N., Kato, T., Hato, F., Mizuno, K., Kobayashi, H., Ishii, M., Kitagawa, S. (2004) Actin reorganization and morphological changes in human neutrophils stimulated by TNF, GM-CSF, and G-CSF: the role of MAP kinases. *Am. J. Physiol. Cell Physiol.* 286: C55-C64.

[33] Hannigan, M., Zhan, L., Li, Z., Ai, Y., Wu, D., Huang, C. K. (2002) Neutrophils lacking phosphoinositide 3-kinase γ show loss of directionality during N-formyl-Met-Leu-Phe-induced chemotaxis. *Proc. Natl. Acad. Sci. USA* 99: 3603-3608.

[34] Kamata, N., Kutsuna, H., Hato, F., Kato, T., Oshitani, N., Arakawa, T., Kitagawa, S. (2004) Activation of human neutrophils by granulocyte colony-stimulating factor, granulocyte-macrophage colony-stimulating factor and tumor necrosis factor-α: the role of phosphatidylinositol 3-kinase. *Int. J. Hematol.* 80: 421-427.

[35] Hannigan, M. O., Zhan, L., Ai, Y., Kotlyarov, A., Gaestel, M., Huang, C. K. (2001) Abnormal migration phenotype of mitogen-activated protein kinase-activated protein kinase $2^{-/-}$ neutrophils in Zigmond chambers containing formyl-methionyl-leucyl-phenylalanine gradients. *J. Immunol.* 167: 3953-3961.

[36] Yen, H., Zhang, Y., Penfold, S., Rollins, B. J. (1997) MCP-1-mediated chemotaxis requires activation of non-overlapping signal transduction pathways. *J. Leukoc. Biol.* 61: 529-532.

[37] Cambien, B., Pomeranz, M., Millet, M. A., Rossi, B., Schmid-Alliana, A. (2001) Signal transduction involved in MCP-1-mediated monocytic transendothelial migration. *Blood* 97: 359-366.

[38] Kintscher, U., Wakino, S., Kim, S., Fleck, E., Hsueh, W. A., Law, R. E. (2001) Angiotensin II induces migration and Pyk2/paxillin phosphorylation of human monocytes. *Hypertension* 37: 587-593.

[39] McGilvray, I. D., Tsai, V., Marshall, J. C., Dackiw, A. P., Rotstein, O. D. (2002) Monocyte adhesion and transmigration induce tissue factor expression: role of the mitogen-activated protein kinases. *Shock* 18:51-57.

[40] Fine, J. S., Byrnes, H. D., Zavodny, P. J., Hipkin, R. W. (2001) Evaluation of signal transduction pathways in chemoattractant-induced human monocyte chemotaxis. *Inflammation* 25: 61-67.

[41] Fields, T. A., Casey, P. J. (1997) Signalling functions and biochemical properties of pertussis toxin-resistant G-proteins. *Biochem. J.* 321: 561-571.

[42] Migeotte, I., Communi, D., Parmentier, M. (2006) Formyl peptide receptors: a promiscuous subfamily of G protein-coupled receptors controlling immune responses. *Cytokine Growth Factor Rev.* 17: 501-519.

[43] Sasaki, T., Kishi, M., Saito, M., Tanaka, T., Higuchi, N., Kominam, E., Katunuma, N., Murachi, T. (1990) Inhibitory effect of di- and tripeptidyl aldehydes on calpains and cathepsins. *J. Enzyme Inhib.* 3: 195-201.

[44] Wang, K. K., Nath, R., Posner, A., Raser, K. J., Buroker-Kilgore, M., Hajimohammadreza, I., Probert, A. W. Jr., Marcoux, F. W., Ye, Q., Takano, E., Hatanaka, M., Maki, M., Caner, H., Collins, J. L., Fergus, A., Lee, K. S., Lunney, E. A., Hays, S. J., Yuen, P. (1996) An alpha-mercaptoacrylic acid derivative is a selective nonpeptide cell-permeable calpain inhibitor and is neuroprotective. *Proc. Natl. Acad. Sci. USA* 93: 6687-6692.

[45] Gao, J. L., Becker, E. L., Freer, R. J., Muthukumaraswamy, N., Murphy, P. M. (1994) A high potency nonformylated peptide agonist for the phagocyte N-formylpeptide chemotactic receptor. *J. Exp. Med.* 180: 2191-2197.

[46] Nanamori, M., Cheng, X., Mei, J., Sang, H., Xuan, Y., Zhou, C., Wang, M. W., Ye, R. D. (2004) A novel nonpeptide ligand for formyl peptide receptor-like 1. *Mol. Pharmacol.* 66: 1213-1222.

[47] Fiore, S., Maddox, J. F., Perez, H. D., Serhan, C. N. (1994) Identification of a human cDNA encoding a functional high affinity lipoxin A4 receptor. *J. Exp. Med.* 180: 253-260.

[48] Neptune, E. R., Bourne, H. R. (1997) Receptors induce chemotaxis by releasing the betagamma subunit of Gi, not by activating Gq or Gs. *Proc. Natl. Acad. Sci. USA* 94: 14489-14494.

[49] Shibata, T., Imaizumi, T., Tamo, W., Matsumiya, T., Kumagai, M., Cui, X.F., Yoshida, H., Takaya, S., Fukuda, I., Satoh, K. (2002) Proteasome inhibitor MG-132 enhances the expression of interleukin-6 in human umbilical vein endothelial cells: Involvement of MAP/ERK kinase. *Immunol. Cell Biol.* 80: 226-230.

[50] Wu, H. M., Wen, H. C., Lin, W. W. (2002) Proteasome inhibitors stimulate interleukin-8 expression via Ras and apoptosis signal-regulating kinase-dependent extracellular signal-related kinase and c-Jun N-terminal kinase activation. *Am. J. Respir. Cell Mol. Biol.* 27: 234-243.

[51] Kwak, H. B., Lee, M. S., Kim, H. S., Cho, H. J., Kim, J. W., Lee, Z. H., Oh, J. (2008) Proteasome inhibitors induce osteoclast survival by activating the Akt pathway. *Biochem. Biophys. Res. Commun.* 377: 1-6.

[52] Chen, D. L., Ping, Y. F., Yu, S. C., Chen, J. H., Yao, X. H., Jiang, X. F., Zhang, H. R., Wang, Q. L., Bian, X. W. (2009) Downregulating FPR restrains xenograft tumors by impairing the angiogenic potential and invasive capability of malignant glioma cells. *Biochem. Biophys. Res. Commun.* 381: 448-452.

[53] Khau, T., Langenbach, S. Y., Schuliga, M., Harris, T., Johnstone, C. N., Anderson, R. L., Stewart, A. G. (2011) Annexin-1 signals mitogen-stimulated breast tumor cell proliferation by activation of the formyl peptide receptors (FPRs) 1 and 2. *FASEB J.* 25: 483-496.

[54] Murphy, P. M., Ozçelik, T., Kenney, R. T., Tiffany, H. L., McDermott, D., Francke, U. (1992) A structural homologue of the N-formyl peptide receptor. Characterization and chromosome mapping of a peptide chemoattractant receptor family. *J. Biol. Chem.* 267: 7637-7643.

[55] Todd, B., Moore, D., Deivanayagam, C. C., Lin, G. D., Chattopadhyay, D., Maki, M., Wang, K. K., Narayana, S. V. (2003) A structural model for the inhibition of calpain by calpastatin: crystal structures of the native domain VI of calpain and its complexes with calpastatin peptide and a small molecule inhibitor. *J. Mol. Biol.* 328: 131-146.

In: Enzymes and Enzyme Activity
Editor: Eva M. Lashinski

ISBN: 978-1-62417-696-8

Chapter VI

ALDH Activity of *Artemia* As a Tool for the Investigation of the Toxicity of Antifouling Paints

***J. Castritsi-Catharios*[1*], *G. Zouganelis*[2], *N. Bourdaniotis*[3] *and S. Koinis*[4]**

[1]Department of Ichthyology and Aquatic Environment, School of Agricultural Sciences, University of Thessaly, Nea Ionia, Volos, Greece

[2]University of Portsmouth, School of Biological Sciences , Portsmouth, United Kingdom.

[3]National and Kapodistrian University of Athens, Department of Biology, Sector of Zoology and Marine Biology, University Campus, Athens, Greece

[4]National and Kapodistrian University of Athens, Department of Chemistry, Laboratory of Inorganic Chemistry, University Campus, Athens, Greece

[*] Corresponding author: Ass. Prof. Dr. J. Castritsi – Catharios, Telephone: +6972234770, Fax: +30 210 6200075, E-mail: cathario@apae.uth.gr & cathario@biol.uoa.gr

Abstract

Leaching of toxic substances from the matrix of antifouling paints affects not only the fouling organisms but also "non-target" biota. *Artemia* nauplii have been found to be suitable test organisms for measurement of acute toxicity effects of certain antifouling paints. The present study addresses the impact of a self polishing copper antifouling paint on the activity of the brine shrimp aldehyde dehydrogenase (ALDH). Total ALDH activities from nauplii exposed to several sublethal concentrations ranging from $(S/V)_{50}$ to $1/4(S/V)_{50}$ were determined. Decreased enzymatic activities were observed in all antifouling paint concentrations tested, ranging from 2.4% inhibition at $1/4(S/V)_{50}$ to 53.9% at $(S/V)_{50}$. The results of this preliminary investigation suggest that ALDH activity of *Artemia* nauplii could be used as a biomarker for the evaluation of toxic activity of antifouling paint on non target organisms.

Keywords: Enzymatic activity, biomarker, copper compounds, toxicity measurement, Artemia nauplii, total aldehyde dehydrogenase

1. Introduction

Antifouling paints are used on a wide range of underwater structures in order to protect them from the development of fouling organisms. The constructions concerned include ship bilges, oil rigs, cofferdams and cages of aquaculture, sonar equipment, underwater pipes and docks (Evans et al., 2000; Boxall et al. 2000; Omae, 2003). While beneficial to the protection of underwater structures from the development of fouling, the antifouling paints have been shown to exhibit toxic effects not only to the fouling organisms but also on "non-target" biota. This fact was fully recognized in the 80's when oyster farming in the bay of Arcachon in France nearly collapsed (Alzieu et al., 1999; Michel and Averty, 1999).

As a result, many investigators tested toxicity effects of antifouling paints on some target and non-target species such as seaweeds *Ulva lactuca* (Grant-Burgess et al., 2003), barnacles *Balanus amphitrite* (Grant-Burgess et al., 2003; Löschau and Krätke, 2005) and brine shrimp *Artemia* nauplii (Bakoulia et al., 2002; Dimitriou et al 2003, Katranitsas et al., 2003; Löschau and Krätke, 2005). *Artemia* nauplii are easily hatched from dry cysts and are available (Sorgeloos et al, 1978).

The crustacean *Artemia* is the main component of zooplankton in natural salt ponds and other saline environments and is suitable as an environmental bioindicator because it can reflect the degree of contamination by trace elements in the aquatic compartment (Petrucci et al, 1995).

The aim of this study is to examine the effect of a self polishing copper-based antifouling agent (SPC) on the enzymatic activity of aldehyde dehydrogenase (ALDH) of freshly hatched larvae of the brine shrimp *Artemia franciscana.*

Although the mean lethal concentration $[L(S/V)_{50}]$ worked out by Persoone and Castritsi-Catharios (1989) on the surface to volume concept, is an important indicator of the toxicity for fouling agents, it could be argued that it does not point out the effects of sublethal concentrations on the physiology of the affected organisms. For this reason, for the present study ALDH activity was selected as a sublethal effect criterion.

ALDH enzymes are important for many physiological, pharmacological and biological processes (Perreira et al.; Yoshida, 1991). ALDH activity is crucial for the regulation of retinoic acid and the neurotransmitter GABA (Yoshida et al., 1992). ALDH is involved in processes of toxicological interest such as alcohol metabolism in vertebrates (Weiner, 1979) and invertebrates (Fry and Saweikis, 2006) and for maintenance of cellular homeostasis by eliminating reactive aldehydes derived from lipid peroxidation (Vasiliou et al., 2000).

In addition these enzymes which are found in most organisms are well characterized and the decrease of their activity can be used to measure the toxic impact of individual chemicals (Keung and Valee, 1992; Winzer and Kohler, 1998; Vasiliou and Marslelos et al., 1989; Vasiliou et al., 2000) or mixtures of them, as in the case of antifouling paint formulation.

2. Materials and Methods

2.1 Antifouling Paint

The commercial copper-based antifouling coating (SPC) "High-Speed Self Polishing Antifouling", produced by the Titan Yate Company was used. This coating is used on high-speed vessels, and contains cuprous oxide (Cu_2O), cuprous thiocyanate (CuSCN) and the booster biocide dichlofluanid as basic active ingredients.

2.2. Artificial Seawater

Artificial seawater of 33‰ salinity made up with Instant Ocean® salts, was used both for the hatching of *Artemia* cysts and for the acute toxicity tests.

2.3. Test Vessels

20 mL PVC test tubes were used to examine the test organisms in the various concentrations of copper.

2.4 Test Organisms

Dehydrated cysts of the *Artemia* Ocean Star International were used. In order to obtain larvae in the Instar I stage, the *Artemia* cysts were incubated for 24 hours at 25 ± 1 °C in a thermostatic chamber with constant aeration and light supply. In all the experiments test organisms of the same age were used.

2.5. Antifouling Paint Toxicity Determination

The toxicity of the antifouling paint was initially determined according to the method of Castritsi-Catharios et al. (2007) also based on the surface to volume concept.

According to this method, different concentrations of the leached toxicants are obtained by adding different volumes of artificial seawater to coated pistons inside specially modified cylinders. Since the area of each painted surface is exactly the same for all the disks (314mm), the concentration of the active element released is decreased by increasing the volume of seawater to which the disks are exposed.

The test organisms were exposed at 25°C for 24h to different concentrations of the active element. The ALDH activity was determined and expressed in terms of painted surface to water volume (S/V) with calculation of mean lethal and sublethal concentrations at different test concentrations: $L(S/V)_{50}$, $1/2(S/V)_{50}$, $1/3(S/V)_{50}$ and $1/4(S/V)_{50}$. For each S/V value, 3 repeats were performed, i.e. a total of 12 samples and 3 controls. At the end of the 24h exposure per sample (12000 in total) 800 test organisms (*Artemia* Instar II-III) were collected with the use of an appropriate filter, after which the living organisms were separated from the dead ones with a Gilson pipet. The organisms were then collected from the filter by pipetting repeatedly up and down Tris-HCl buffer and homogenized with an appropriate homogenizer. The homogenized solutions were kept on ice.

2.6. Enzyme Assays-Photometric Determination

The determination of enzymatic activity was carried out according to the method of Vassiliou and Marselos (1989). The photometric enzymatic assays of the thermostated (25°*C*) samples, were measured in 1 cm pathlength quartz cuvettes in a Cary 3E (Varian) UV-Vis Spectrophotometer, equipped with Varian Temperature Controller, at 340 nm for a time course of 300 sec.

For the determination of the total ALDH activity the following solutions were used: 225 μL 0.1M Na-pyrophosphate, pH 8.0 (4.46 g/L) (Merck), 15 μL 20 mM pyrazole (0.1360 g/100 mL) (Merck), 15 μL Solution C [20 mM β-nicotinamide adenine dinucleotide (NAD, 14 mg/mL) (Serva), 30 μL 100 mM propionaldehyde (7.26 mL/100mL solution) (Merck), 0,25 M sucrose and 25 mM Tris HCl buffer, pH 7.0 (Serva).

Furthermore, 15 μL *Artemia* extracts were included from the biotest as described in section 2.5. These were solutions ranging from 0.1 to 1 units/mL ALDH (Sigma A-6338) at 0.1 units/mL increments. In the assay, a blank was included with 15 μL solution E (homogenizing buffer) instead of *Artemia* extract.

A standard activity assay curve for pure ALDH (Sigma) was constructed using the same experimental conditions. The enzymatic activity of Instar II-III *Artemia* nauplii extracts expressed in nmol NAD(H)/min·mg protein was calculated and correlated with the values from the standard activity assay.

2.7. Copper Measurement

The concentration of copper in the experimental solutions was determined by Direct Aspiration Atomic Absorption Spectrometry (DAAAS) according to the ASTM D1688-02 standard at 324.8 nm.

2.8. Statistical Analysis

The statistical analysis of the results was done with the use of the statistical package ORIGIN (Version 7.5).

2.9. Bioinformatics Analysis of *Artemia* ALDH

The partial protein sequence corresponding to *Artemia* ALDH derived from mRNA found in Genebank (Accesion Number: EF203469) was compared with deposited Genebank database sequences with the use of BLAST (Altschul et al., 1990). Sequences with high similarity from diverse vertebrate species and invertebrate species were retrieved and functionally annotated with the use of the Conserved Domain Database analysis programs (Marchler-Bauer et al., 2010). The sequences illustrated here correspond to aldehyde dehydrogenases class III from *Rattus norvegicus* (Accession Number: NP_001006999.1), *Danio rerio* (Accesion number: MN_001004658), *Ciona intestinalis* (Accession number: XM_002124113) and *Nasonia vitripenis* (Accesion number: XM_003426876).

3. Results

The experimental concentrations for the determination of the ALDH activity are given in Table 1. The values of ALDH enzymatic activity for each experimental concentration are shown in Table 2 for both the *Artemia* extracts and the blank. The statistical analysis of the results (one way Anova) showed significant differences at each test concentration at the 95 % confidence level between the values of the activities.

Table 1. Experimental concentrations for the determination ALDH activity

	$1/4(S/V)_{50}$	$1/3(S/V)_{50}$	$1/2(S/V)_{50}$	$(S/V)_{50}$	CONTROLS
S/V (mm^2/mL)	5.016	6.689	10.032	20.065	0
mL salt water	62.6	46.94	31.3	15.65	-

Table 2. Mean enzymatic activity values for every concentration

S/V (mm^2/mL)	Mean enzymic activity values (nmol NAD(H)/min.mg protein) ±SD	% Inhibition	Copper oxide concentration in mg/l
Blank (0)	113,831±0,598	0	0
$1/4(S/V)_{50}$ (5.016)	112,216±1,305	1.4%	0.92
$1/3(S/V)_{50}$ (6.689)	108,529±1,167	4.7%	1.11
$1/2(S/V)_{50}$(10.032)	96,948±4,853	14.8%	1.48
$(S/V)_{50}$ (20.065)	56,796±12,693	50.1%	1.81

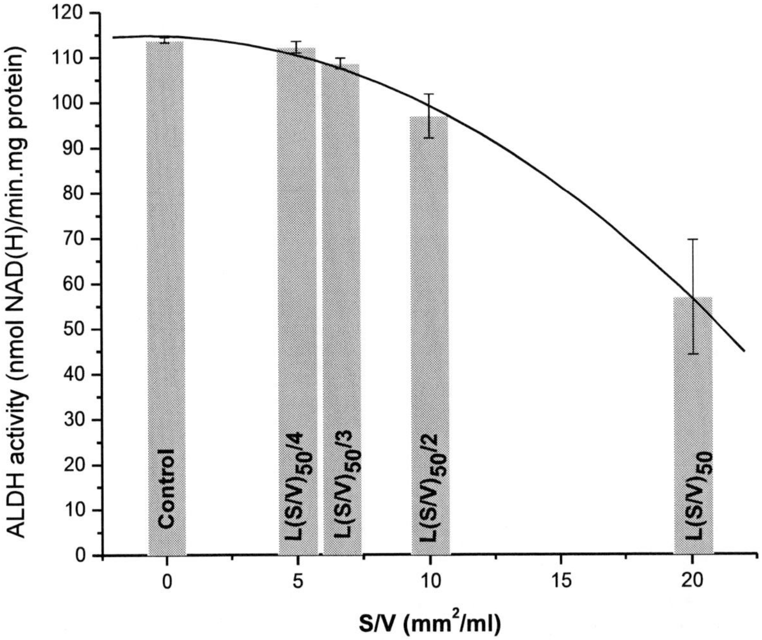

Figure 1. Activity change of ALDH in comparison to the increase of antifouling agent concentration.

Furthermore, we applied a series of t-tests to show the significance in value differences. The control sample value exhibits a significant statistical difference from all S/V values as follows: $1/4(S/V)_{50}$ ($P=0.00877 \leq 0.05$), $1/3(S/V)_{50}$ (P= 0,00177<0.05), $1/2(S/V)_{50}$ (P= 0,00350<0.05) and $(S/V)_{50}$ (P= 0,00003<0.05).

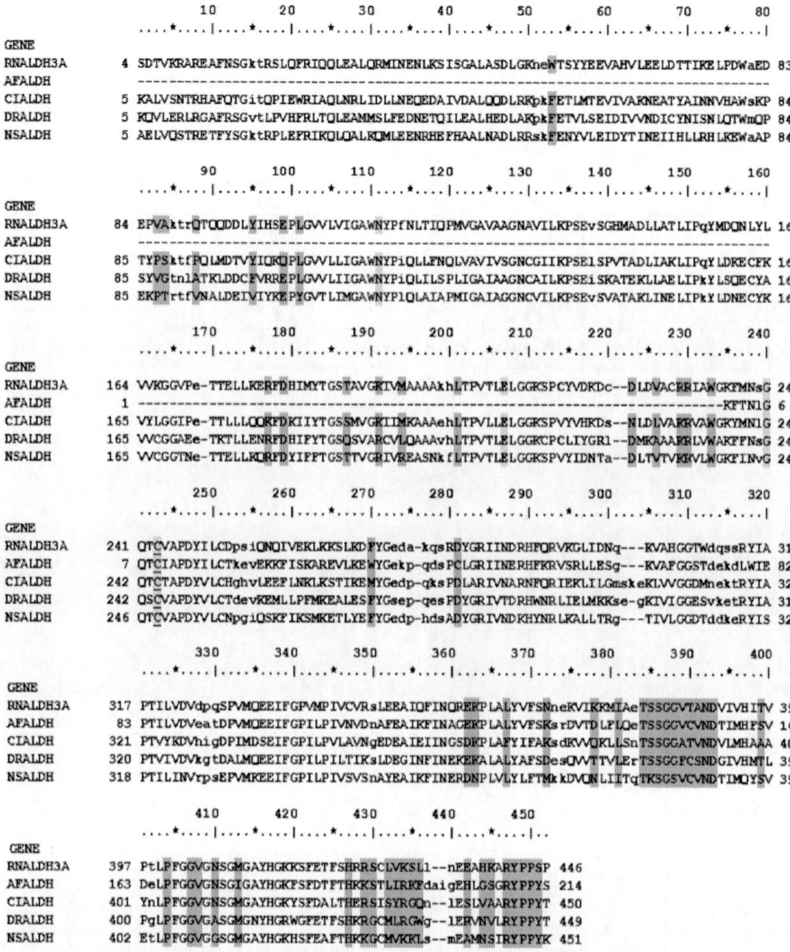

Figure 2: Comparison of structural features between ALDH and with the *Artemia* partial ALDH sequence (Accesion Number: EF203469). Sequences compared here are from *Rattus norvegicus* (RNALDH3A, Accession Number: NP_001006999.1), *Ciona intestinalis* (CIALDH, Accesion number: MN_001004658), *Danio rerio* (DRALDH, Accession number: XM_002124113), and *Nasonia sp.* (NSALDH, Accesion number: XM_003426876). The aminoacids in the catalytic site are highlighted with yellow with the binding cysteine (C̲) undelined, the homodimeric sites are highlighted with grey and the NADP binding sites are shown in blue.

Additionally, t-test shows that $L(S/V)_{50}$ values exhibit significant differences as follows: $1/4(S/V)_{50}$, $1/3(S/V)_{50}$ (P=0,01773<0.05), $1/4(S/V)_{50}$, $1/2(S/V)_{50}$ (P=0,00649<0.05) and $1/3(S/V)_{50}$, $1/2(S/V)_{50}$ (P= 0,01540<0.05)

(Figure 1). Table 2 also contains the date of copper concentrations of each test concentration and shows that the total ALDH activity inhibition increases with the increase of copper concentration.

The partial protein sequence corresponding to *Artemia* ALDH is aligned and compared with protein sequences found in vertebrate and invertebrate species and the structural features are demonstrated in Figure 2.

Conclusion

The activity of ALDH from the brine shrimp *Artemia* decreases significantly with the concentration increase of the active compound contained in the antifouling paint. The activity of the enzyme is inhibited by 2.4% at the lowest test concentration $1/4(S/V)_{50}$ up to 53.9% at the highest test concentration $(S/V)_{50}$.

ALDH has already been previously used as a bioindicator for the monitoring of pollution in marine systems (Nesto et al., 2004). In a study on mussels in the Venice lagoon (Nasci et al., 2004), inhibition was observed in the ALDH activity as a result of agricultural, industrial and domestic pollution (important sources of heavy metals such as copper). The industrial area of Marghera in Italy, where the samples for the above study were taken, includes among other industrial activities, many shipyards. Another study with samples from the same region showed decreased activity in samples collected in areas with heavy pollution. In contrast, in the species *Limanda limanda* it was found that total ALDH activity is increased in samples found in heavily polluted areas. It is important to note that the detection of ALDH activity depends upon the substrate used. In this study propionaldehyde was used as a substrate which is very sensitive in comparison to other substrates (Winzer and Kohler, 1998). The effect of copper to the activity of ALDH has been studied in hamsters provided with food rich or poor in copper content. The results here were controversial because lack of copper in food leads to increase in enzymatic activity in female individuals and decrease in male individuals. These results were affected by the fact that the evaluation was done in 66 days old animals that might not have been reproductively mature and therefore the expression of enzymes might have been affected by reproductive hormones (Millo and Werman, 2002).

The inhibition observed in ALDH activity in the present study might be attributed to both copper ions and complementary biocide dichlofluanid that are released from the antifouling paint. As shown in Figure 1, we have found in our samples that the total copper concentration increases in higher $(S/V)_{50}$ fractions, while ALDH activity is inhibited. The most abundant form of dissolved copper in marine environments is Cu^{+2} (Hall and Anderson, 1999). This can be produced from the insoluble cuprous oxide (Cu_2O) by oxidation and/or disproportionation giving rise to the formation of soluble Cu^{+2} species (Cotton and Wilkinson, 1988). When copper ions enter into the intra-cellular matrix they are reduced to Cu^{+1} which shows an increased affinity for coordination with thiolato-groups, such as the ones derived from –SH enzyme groups (Variengo et al., 1996). ALDH contains SH-groups (7.5 mole –SH/mole enzyme) (De Master et al., 1997). The coordination of copper with –SH groups induces changes in their conformation and this way prevents the conjunction of other ions (Trieff, 1980). A similar inhibitory mechanism has been proposed for other ALDH inhibitors (aldehydes such as acroleine or ketones) (Kikonyogo et al., 1999). Additionally, the importance of SH-groups in aldehyde dehydrogenase activity has been demonstrated with SH-blocking experiments. Konoplitskaia et al. (1984) showed that SH-group blockers such as tetraethylthiuram disulphide (TETD), 5,5'-dithiobisnitrobenzoic acid (DTNB), p-chloromercurybenzoate (p-ChMB) and N-ethylmaleimide (NEM) reduce rat liver ALDH activity up to 50%. Dichlofluanid is a known "SH-blocker" (Dittrich and Issinger, 1969). The biological activity of dichlofluanid is presumed to be localized in the trihalogeno-methylthio part of the molecule (Kühle et al., 1964). Lemperle et al. (1973) showed that yeast alcohol dehydrogenase and glycerinaldehyde-3-phosphate dehydrogenase which contain essential SH-groups are inhibited in vivo and/or in vitro by dichlofluanid. It is therefore expected that dichlofluanid released from the antifouling paint contributes to the inhibition of *Artemia* ALDH.

In order to assess the effect of copper and dichlofluanid on the SH-groups of *Artemia* ALDH, a preliminary structural analysis and comparison of the brine shrimp enzyme with diverse homologs of vertebrate and invertebrate species was undertaken.

As demonstrated in Figure 2, a remarkable conservation in the catalytic site, homodimeric site and NAD sites is evident. As it appears there are no signifficant -SH groups that affect the enzyme function other than Cysteine-8 of the partial sequence which corresponds to an essential catalytic domain. Coordination of Cysteine-8 with Copper and/or blocking of this site with dichlofluanid may disrupt the formation of thiohemiacetal intermediates and the hydride transfer for the formation of NADH (Wymore et al., 2004).

Althought the inhibitory effect of antifouling formulation components appears possible, an overall effect on the expression of particular classes of *Artemia* enzymes cannot be excluded. Toxins derived from diatoms have been shown to modify expression of key metabolic enzymes in copepods (Lauritano et al., 2011). Copper has be shown to reduce the expression levels of alcohol dehydrogenase in *Chironomus riparius* larvae (Park and Kwak, 2009). Future studies will clarify the interplay between inhibitory effects and regulation of expression of ALDH in *Artemia* as response to the exposure of antifouling components.

To our knowledge this is the first time that ALDH activity in *Artemia* nauplii is measured and the findings of this preliminary study indicate that ALDH might be an interesting bioindicator for measuring toxicity stress from antifouling agents on a non-target organism such as *Artemia*. In addition, the measurement of ALDH activity in natural populations of *Artemia* can also be used as a bioindicator for a rapid estimation of heavy metals contamination level in certain ecosystems such us solar saltworks and coastal brine ponds, where populations of this crustacean species occur. The sensitivity of different *Artemia* strains in several developmental stages should also be investigated with several antifouling paints of the new generation.

Acknowledgments

We would like to express our gratitude to Professor V. Vasiliou for the supply of the work protocol.

References

[1] Altschul, S. F.; Gish, W.; Miller, W.; Myers, E., W.; Lipman D. J. (1990). Basic local alignment search tool. *Journal of Molecular Biology*, 215, 403-10.

[2] Alzieu, C. (1998). Tributyltin: Case study of chronic contaminant in the coastal environment. *Ocean Coast Management*, 40, 23-36.

[3] Bakoulia, P.; Marcouli, P.; Iliopoulou-Georgudaki, J. (2002). Acute Toxicity of TBT and IRGAROL in *Artemia Salina*. *Journal of Toxicology*, 21, 231-233.

[4] Boxall, A.B.A.; Comber, S.D.; Conrad, A.U.; Howcroft, J.; Zaman, N.,(2000). Inputs, monitoring and fate modeling of antifouling biocides in UK estuaries. *Marine Pollution Bulletin* 40, 898-905.

[5] Castritsi – Catharios, I.; Bourdaniotis N.; Persoone P. (2007). A new simple method with high precision for determining the toxicity of antifouling paints on brine shrimp larvae (*Artemia*): first results. *Chemosphere*, Vol. 67, Iss. 6, 1127 - 1132.

[6] Cotton, F.A. & Wilkinson, G.(1988). Advanced Inorganic Chemistry, 5th edition. New York , Wiley-Interscience.

[7] DeMaster, E.G.; Redfern, B.; Quast, B.J.; Dahlseid, T.; Nagasawa, H.T. (1997). Mechanism for the inhibition of aldehyde dehydrogenase by nitric oxide. *Alcohol* 14, 181-189.

[8] Dimitriou, P.; Castritsi – Catharios, J.; Miliou, E. (2003). Acute toxicity effects of Tributyltin Chloride and Triphenylin Chloride on gilthead seabream Sparus aurata L. embryos. *Ecotoxicology and Environmental Safety*, 54, 30-35.

[9] Dittrich, H. H.; Issinger, O.G. (1969). Zum Wirkungsmechanismus von Dichlofluanid (N,N-Dimethyl-N'-phenyl-[N'-Fluordichlormethylthio]-Sulfamid). A. Pflanzenkr. Pflanzenschutz, 76, 651-663.

[10] Evans, S.M.; Kerrigan, E.; Palmer, N. (2000). Causes of imposex in the dogwhelk *Nucella lapillus* (L.) and its use as a biological indicator of tributyltin contamination. *Marine Pollution Bulletin*, 40, 212-219.

[11] Fry, J. D.; Saweikis, M. (2006). Aldehyde dehydrogenase is essential for both adult and larval ethanol resistance in *Drosophila melanogaster*. *Genetics Research*, 87, 87-92.

[12] Grant-Burgess, J.; Boyd, K. G.; Armstrong, E.; Zhong, J.; Yan, L.; Berggren, M.; May, U.; Pisacane, T.; Granmo, A.; Adams, D. R. (2003). The development of a Marine Natural Product-based Antifouling Paint. *Biofouling*, 19, 197-205.

[13] Hall, L.W.; Anderson R.D. (1999). A deterministic ecological risk assessment for copper in European saltwater environments. *Marine Pollution Bulletin* 38, 207-218.

[14] Katranitsas, A.; Castritsi-Catharios, J.; Persoone, G. (2003). The effects of a copper-based antifouling paint on mortality and enzymatic activity of a non-target marine organism. *Marine Pollution Bulletin*, 46, 1491-1494.

[15] Keung, W.; Valee, B. L.; (1992). Daidzin: a potent, selective inhibitor of human mitochondrial aldehyde dehydrogenase. *Proceeding of National Academy of Science U.S.A* 90, 1247-1251.

[16] Kikonyogo, A.; Abriola, D.P.; Dryjanski, M.; Pietruszko, R. (1999). Mechanism of inhibition of Aldehyde dehydrogenase by citral, a retinoid antagonist. *European Journal of Biochemistry*, 262, 704-712.

[17] Konoplitskaia, K. L.; Kuz'mina, G.I.; Grigor'eva, M.V.; Pozniakova, T.N. (1984). Effect of SH-reagents on aldehyde dehydrogenase activity of the rat liver. *Ukrainskiĭ biokhimicheskiĭ zhurnal*, 56, 628-633.

[18] Kühle, E.; Klauke, E.; Grewe, F. (1964).. Fluordichlormethylthio-Verbindungen und ihre Verwendung im Pflanzenschutz. *Angewandte chemie*, 7, 807-816.

[19] Lemperle, E.; Kerner, E., Strecker, H. and Waibel, A. 1973. Wirkstoffrückstände und Gärbeeinflussungen nach Anwendung von Fungiziden im Weinbau. *Deutsche Lebensmittel-Rundschau* 69, 313-322.

[20] Lauritano, C., Borra, M.; Carotenuto, Y., Biffali, E.; Miralto, A.; Procaccini, G.; Ianora, A. (2011). Molecular evidence of the toxic effects of diatom diets on gene expression patterns in copepods. *PLoSONE*, 6, e26650.

[21] Löschau, M.; Krätke, R. (2005). Efficacy and toxicity of self polluting biocide-free antifouling paints. *Environmental Pollution*, 138, 260-267.

[22] Marchler-Bauer, A.; Lu, S.; Anderson, J.B.; Chitsaz, F.; Derbyshire, M. K.; DeWeese-Scott, C.; Fong, J.H.; Geer, L.Y.; Geer, R. C.; Gonzales N. R.; Gwadz M.; Hurwitz, D. I.; Jackson, J. D.; Ke, Z.; Lanczycki, C.J.; Lu, F.; Marchler, G.H.; Mullokandov, M.; Omelchenko, M. V.; Robertson, C. L.; Song ,J. S.; Thanki, N.;

[23] Yamashita, R. A.; Zhang, D.; Zhang, N.; Zheng, C.;Bryant S. H. (2010). CDD: a Conserved Domain Database for the functional annotation of proteins. Nucleic Acids Research, doi: 10.1093/nar/gkq1189.

[24] Michel P.; Averty, B. (1999). Contamination of French Coastal Waters by Organotin Compounds: 1997 Update. Marine Pollution. *Bulletin*, 38, 268-275.

[25] Millo, H.; Werman, M.J., (2002). Hepatic fructose-metabolizing enzymes and related metabolites: Role of dietary copper and gender. *Journal of Nutritional Biochemistry*, 11, 374-381.

[26] Nasci, C.; Nesto, N.; Monteduro, R.A.; Da Ros, L. (2002). Field application of biochemical markers and a physiological index in the mussel, *Mytilus galloprovincialis*: transplantation and biomonitoring studies in the Lagoon of Venice (NE Italy). *Marine Environmental Research*, 54, 811-816.

[27] Nesto, N.; Bertoldo, M.; Nasci, C.; Da Ros, L. (2004). Spatial and temporal variation of biomarkers in mussels (*Mytilus galloprovincialis*) from the Lagoon of Venice, Italy. *Marine Environmental Research*, 58, 287-291.

[28] Omae, I. (2003). General Aspects of Tin-Free Antifouling Paints. *Chemical Reviews*, 103, 3431-3448.

[29] Park, K.; Kwak, I. (2009). Characterization and expression of *Chironomus riparius* alcohol dehydrogenase gene under heavy metal stress. *Journal of Environmental Toxicology*, 24, 107-117.

[30] Pereira, F.; Rosenmann, E.; Nylen, E.; Kaufman, M.; Pinsky, L.; Wrogemann, K. (1991). The 56 kDa androgen binding protein is an aldehyde dehydrogenase. *Biochemical and Biophysical Research Communications*, 175, 831-838.

[31] Petrucci, F.; Caimi, S.; Mura, G.; Caroli, S. (1995). *Artemia* as a bioindicator of environmental contamination by trace elements. *Microchemical Journal*, 51, 1-2, 181-186

[32] Persoone, G.; Castritsi-Catharios, J. (1989). A simple bioassay with Artemia larvae to determine the acute toxicity of antifouling paints. *Water Research*, 23, 893-897.

[33] Sorgeloos, P.; Remiche – Van der Wielen, C.; Persoone, G. (1978). The use of *Artemia* nauplii for toxicity test : a critical analysis. *Ecotoxicology and Environmental Safety*, *2*,: 249-255

[34] Trieff, N.M. (1980). Toxicity of heavy metals, oils and other organics on *Artemia*. In: Persoone, G., Sorgeloos, P., Roels, O., Jaspers, E., (Eds.), The brine shrimp *Artemia*: Morphology, Genetics, Radiobiology, Toxicology,1. Welteren, Universal Press, pp. 253-262.

[35] Vasiliou, V.; Marselos, N. (1989). Tissue distribution of inducible ALDH activity in the rat after treatment with Phenobarbital or methylcholanthrene. *Pharmacology and Toxicology*, 64, 39-42.

[36] Vasiliou, V.; Pappa, A.; Petersen, D.R. (2000). Role of aldehyde dehydrogenases in endogenous and xenobiotic metabolism. *Chemico - Biological Interactions*, 129,1-19.

[37] Viarengo, A.; Pertica, M.; Mancinelli, G.; Burlando, B.; Canesi, L.; Orunesu, M. (1996). In vivo effects of copper on the calcium homeostasis mechanism of mussel gill cell plasma membranes. *Comparative Biochemistry and Physiology*, 113C, 421-425.

[38] Weiner, H. (1979). Acetaldehyde metabolism. In Biochemistry and Pharmacology of Ethanol, Vol. 1 Majchrowicz E., & Noble E., P. (Eds.), pp.125-144.

[39] Wymore, T.; Hempel, J.; Cho, S.S.; MacKerell, A.D.; Nicholas Jr., H.B.; Deefield, D.W. (2004). Molecular recognition of aldehydes by aldehyde dehydrogenase and mechanism of nuclophile activation. *Proteins:Structure, Function and Bioinformatics*, 57, 758-771.

[40] Winzer, K.; Kohler, A. (1998). Aldehyde dehydrogenase and Glucose-6-phosphate dehydrogenase as Markers for enzyme-altered foci the liver of Dab (*Limanda Limanda* L.). *Marine Environmental Research*, 46, 215-219.

[41] Yoshida, A; Hsu, LC; Yasunami M. (1991). Genetics of human alcohol-metabolizing enzymes. *Progress in nucleic acid research and molecular biology*, 40, 255–287.

[42] Yoshida, A.; Hsu, LC; Dave′, V. (1992). Retinal oxidation activity and biological role of human cytosolic aldehyde dehydrogenase. *Enzyme*, 46, 239–244.

Index

A

D

E

F

G

H

I

J

K

L

M

N

Q

R

S